苯二胺衍生碳点的表面态调控策略及其应用研究

李天泽　董媛媛　著

图书在版编目（CIP）数据

苯二胺衍生碳点的表面态调控策略及其应用研究 / 李天泽，董媛媛著. -- 哈尔滨：黑龙江科学技术出版社，2023.2

ISBN 978-7-5719-1748-7

Ⅰ. ①苯… Ⅱ. ①李… ②董… Ⅲ. ①荧光 - 纳米材料 - 研究 Ⅳ. ①TB383

中国国家版本馆 CIP 数据核字(2023)第 025469 号

苯二胺衍生碳点的表面态调控策略及其应用研究
BENERAN YANSHENG TANDIAN DE BIAOMIANTAI TIAOKONG CELUE JIQI YINGYONG YANJIU
李天泽 董媛媛 著

责任编辑 闫海波
封面设计 林 子
出 版 黑龙江科学技术出版社
地址：哈尔滨市南岗区公安街 70-2 号 邮编：150007
电话：（0451）53642106 传真：（0451）53642143
网址：www.lkcbs.cn
发 行 全国新华书店
印 刷 哈尔滨博奇印刷有限公司
开 本 787 mm×1092 mm 1/16
印 张 13.25
字 数 250 千字
版 次 2023 年 2 月第 1 版
印 次 2023 年 2 月第 1 次印刷
书 号 ISBN 978-7-5719-1748-7
定 价 68.00 元

前言

碳点(CDs)作为碳基纳米材料领域的一颗冉冉升起的新星,自被发现以来就备受关注。苯二胺[$C_6H_4(NH_2)_2$]具有高活性氨基,在合适的条件下苯二胺可以通过氧化/聚合反应生成各种氧化产物,这些氧化产物可以进一步交联、碳化或聚合得到 CDs。近年来,苯二胺衍生的 CDs 因其在荧光发射方面的突出特性、优异的生物相容性和低毒性而受到广泛关注和研究。与此同时,CDs 表面具有的官能团种类和数量决定了 CDs 的表面态,而不同的表面态对 CDs 的性质和应用具有重要的影响。因此,开发制备不同表面态 CDs 的方法,对拓宽制备 CDs 的思路、获得性能多样的 CDs 以满足后续实际分析需求具有重要意义。本学术专著以三种苯二胺同分异构体中的邻苯二胺(o-PD)为碳源,从分子融合角度出发,通过反应溶剂参与的分子融合法、室温氧化融合法以及氟掺杂的表面态调控法等方法制备了一系列不同表面态的 CDs,并系统地研究了 CDs 的结构和形成机理,解释了表面态与 CDs 发光机理之间的关系,最后根据不同表面态赋予 CDs 的特性,将其应用于不同的检测领域。具体研究内容如下:

1. 本学术专著第一章绪论简要概括了 CDs 的制备方法、光学性质,介绍了几种经典的调控 CDs 表面态/结构的方法(包括:化学掺杂、溶剂效应和表面修饰),总结了苯二胺衍生 CDs 的最新研究进展,讨论了一系列为实现更好的荧光性能和更高级功能开发的不同的制备方法,并根据制备温度的差异对 CDs 的形成机理进行了讨论。对于 CDs 的光学特性,从三个角度进行阐述,包括:吸收、荧光发射和荧光机理。此外,对 CDs 在光学、能源和生物等领域的应用进行了概述。

2. 第二章通过反应溶剂参与的分子融合法制备了不同表面态的 CDs。其中,以 o-PD 和甲酰胺为碳源和反应溶剂制备的 FCDs,不仅具有从碳源和反应溶剂继承的官能团,而且还具有大量通过席夫碱反应产生的 C=N 基团。与其他采用分子融合法制备的

CDs 相比，FCDs 特有的结构和表面态能使其具有良好的“软碱”性能，对 Ag^+ 具有良好的选择性响应，并对硬酸型金属离子呈现出了极强的耐受性。以 FCDs 为荧光探针，基于荧光“关闭”方式建立了 Ag^+ 的荧光检测方法，检出限为 0.019 μmol · L^{-1}。同时，基于 Ag^+ 与半胱氨酸（Cys）之间的相互作用，FCDs/Ag^+ 体系可以通过荧光“开启”方式实现 Cys 的定量分析，检出限为 0.015 μmol · L^{-1}。此外，FCDs 还显示出了良好的生物相容性和较低的细胞毒性，因此可作为荧光探针用于细胞成像。本研究中基于反应溶剂参与的分子融合法的技术路线可以为新型碳基纳米材料的设计以及 CDs 的结构/表面态与性质关系的优化提供新的思路。

3. 随着对 CDs 研究的深入，CDs 的制备过程在向绿色、环保的方向发展。本学术专著第三章基于室温氧化融合法绿色制备了不同表面态 CDs。研究中以 o－PD 和对苯二酚（HQ）为碳源，在室温下通过氧化/聚合和席夫碱反应，无须加入氧化剂的条件下，获得了 CDs 粗产物。基于粗产物组分的极性差异，采用硅胶色谱柱分离后，得到两种 CDs（YCDs 和 GCDs）。不同的表面态赋予这两种 CDs 不同的特性。YCDs 表面富含 $-NO_2$ 和 $-OH$ 基团使其具有较窄的带隙，从而导致其荧光发射峰红移至 555 nm。基于内滤效应，将 YCDs 作为探针用于有毒污染物对硝基苯酚（p－NP）的定量分析，检出限为 0.08 μmol · L^{-1}。GCDs 表面具有丰富的 $-NH_2$ 基团，可以与 H_2O/D_2O 相互作用。由于 H_2O 振荡对 GCDs 的淬灭效果高于 D_2O 振荡的淬灭效果，因此可以将 GCDs 用于 D_2O 中 H_2O 含量的检测，检出限为体积分数 0.17。

4. 第四章通过氟掺杂的表面态调控法，以 o－PD、对苯醌（p－BQ）以及 o－PD 衍生物 4－氟－1，2－苯二胺为碳源，在室温下通过不同的氟掺杂路径制备了具有不同氟含量的 o－PD 衍生 CDs（FCDs1 和 FCDs2）。研究发现，氟掺杂引起的带隙变窄会使制备的两种氟掺杂 CDs 的荧光都发生明显红移。基于元素稀释效应和氢键的形成，FCDs1 中掺入少量氟可以缓解聚集诱导荧光淬灭现象的发生，使 FCDs1 表现出固态荧光特性，并可将其用于潜指纹（LFPs）的高分辨率显现。FCDs1 着色后的指纹显示出清晰的线条和明显的细节，甚至可以清楚地观察到汗孔。具有更高氟含量的 FCDs2 展现了 70 nm 的显著荧光红移。基于内滤效应，将 FCDs2 作为荧光传感器用于钴胺素（CBL）的定量检测，检出限为 0.15 μmol · L^{-1}。

5. 第五章采用溶剂热法，以 o－PD 和乙醇为原料，制备了黄色发光的氮掺杂 CDs（YNCDs）（$\lambda_{ex}/\lambda_{em}$ = 410/555 nm）。基于内滤效应，姜黄素可以有效抑制 YNCDs 的荧光发射。因此，以 YNCDs 为纳米探针，建立了检测姜黄素的“开关”传感器平台，其检出限（3S/N）为 0.01 μmol · L^{-1}。此外，该传感器平台实现了对咖喱粉、人尿、人血清等真实样品中姜黄素的高选择性、高灵敏度检测,准确性和回收率也令人满意。进一步地,YNCDs 优异的光学性能还可作为隐形油墨应用于信息存储和防伪等领域。

6. 第六章采用水热法,以 o-PD 为原料,制备了强黄光 CDs(SYCDs)。结晶紫是一种危险染料,对环境和人类健康构成严重威胁。这促使我们开发一种简便的方法来对结晶紫进行灵敏的检测。结晶紫可以有效抑制 SYCDs 的荧光发射。因此,以 SYCDs 为纳米探针,建立了检测结晶紫的传感器平台,在 0.02 ~ 15.00 $\mu mol \cdot L^{-1}$ 范围内线性良好,其检出限(3S/N)为 0.006 $\mu mol \cdot L^{-1}$。通过详细的研究,提出了内滤效应是检测机理的结论。此外,该传感器平台实现了鱼组织真实样品中结晶紫的高选择性、高灵敏度检测,准确性和回收率也令人满意。

7. 第七章总结了本学术专著的相关研究内容,并对 CDs 的制备及其未来的发展进行了展望。

感谢黑龙江省自然科学基金(项目号:LH2022B020)、黑龙江省省属高等学校基本科研业务费科研项目(项目号:2021BJ03;2021BJ02)对本学术专著出版的支持。在本书撰写过程中,作者阅读了国内外大量的相关文献,在此谨向文献作者表示衷心的感谢,并感谢所有人的支持和鼓励。本学术专著由黑龙江工程学院李天泽和黑龙江工程学院董媛媛担任著者。具体分工如下:李天泽负责第二、三、四、七章(共计 12.3 万字);董媛媛负责第一、五、六章(共计 12.7 万字)。全书由李天泽统稿完成。由于著者的经验和水平所限,书中不妥之处,诚请专家和读者批评指正。

目录

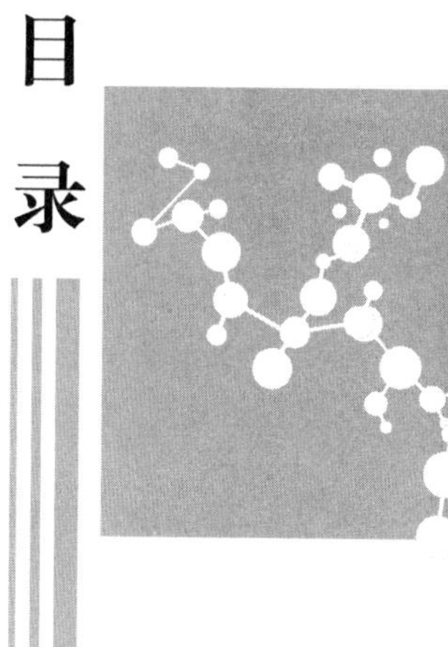

第1章
绪　论

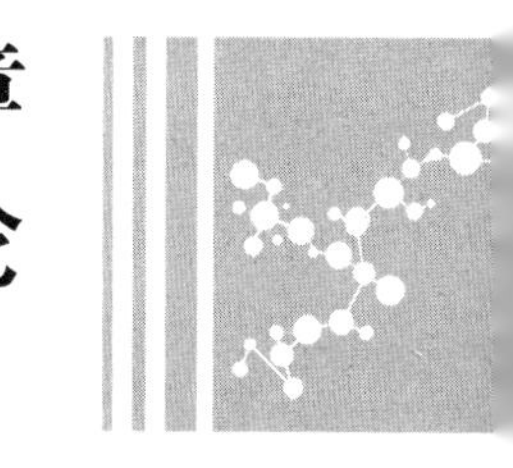

苯二胺衍生碳点的表面态调控策略及其
应用研究

1.1 碳点的概述

碳基材料在材料科学的发展中起着重要作用。从传统工业碳(例如:活性炭和炭黑)到新工业碳(例如:碳纤维和石墨)和新型碳纳米材料(例如:石墨烯和碳纳米管),碳基材料的环境友好性、储量丰富以及无毒性等优点,使其在化学、材料和其他跨学科领域的基础研究和应用中有着举足轻重的地位。然而,传统碳材料缺乏适当的带隙,使其难以作为有效的荧光材料在发光领域有所建树。碳点(CDs)作为碳材料家族中的新星,它的出现填补了碳纳米材料在荧光材料领域应用的空缺。另外,其出色的可调光致发光、高量子产率、低毒、小体积、可观的生物相容性和低成本等特点,也使其成为传统重金属量子点的替代品,用于生物医学、催化、光电元件、传感检测和防伪涂层等重要领域[1-9]。

CDs 是一种准零维碳基纳米材料,荧光是它们的固有属性。21 世纪初,Scrivens 团队在纯化单壁碳纳米管中偶然得到了一种具有荧光性能的碳基纳米颗粒[10]。自此,开启了碳基纳米材料一扇新的大门,而此时并未将其命名为 CDs。2006 年,Sun 及其同事首次将通过激光烧蚀碳靶而获得的纳米级碳颗粒命名为 CDs,而此时这些表面钝化的 CDs 的量子产率仅为 10%[11]。量子产率低且制备过程复杂限制了更多种类 CDs 的开发和应用。直到 2013 年,Yang 课题组通过乙二胺和柠檬酸之间的缩合和碳化反应,制备了聚合物类 CDs,其量子产率高达 80%,这使其既可以作为油墨用于印刷,也可以用于防伪。同时,该 CDs 还可以作为生物传感器用于定量检测生物体系中的 Fe^{3+} 含量[12]。自此,CDs 简便的制备方法、高量子产率、低毒性和高抗光漂白性引起广泛关注和研究热潮,研究人员也开发了不同的方法和技术来追求制备高性能 CDs,并取得了许多重大突破,其中包括:多色/深红色/近红外发射 CDs、双/多光子荧光 CDs、手性 CDs、室温磷光 CDs 和热激活延迟荧光 CDs 等[2,13-25]。

1.1.1 碳点的结构

CDs 的结构由 sp^2/sp^3 碳和基于氧/氮的基团或聚合物组成。如图 1-1 所示,根据不同形成机理、微观/纳米结构和性能,CDs 可以被分为三种类型:石墨烯量子点(GQDs)、碳量子点(CQDs)和碳聚合物点(CPDs),而这三者之间可以通过改变石墨烯层和碳化度而建立联系[26]。GQDs 具有一层或多层石墨结构,在表面/边缘或层间缺陷内通过化学基团连接[27-29]。GQDs 具有明显的石墨烯晶格,通常是通过氧化物切割较大的石墨化碳材料的方法得到的[27-30]。GQDs 表现出各向异性,其横向尺寸大于其高度。而与之相比,CQDs 和 CPDs 是球形的,通常是由小分子、聚合物或生物质以"自下而上"的方法通过组装、聚合、交联和碳化而制得的[21-23]。CQDs 被定义为是一种表面连接了许多基团的多层石墨结构,CPDs 则是衍生自线性聚合物或单体的聚集或交联的聚合物纳米颗粒。值得注意的是,CPDs 这个概念是于 2018 年根据其形成过程、结构和荧光机理提出的[23,31]。它具有特殊的"核-壳"纳米结构,"核"包含高度脱水的交联聚合物框架或轻微的石墨化程度碳,"壳"具有丰富的官能团或聚合物链[12,32],这使 CPDs 表现出更高的稳定性、更好的相容性、更易于表面官能化等特点。

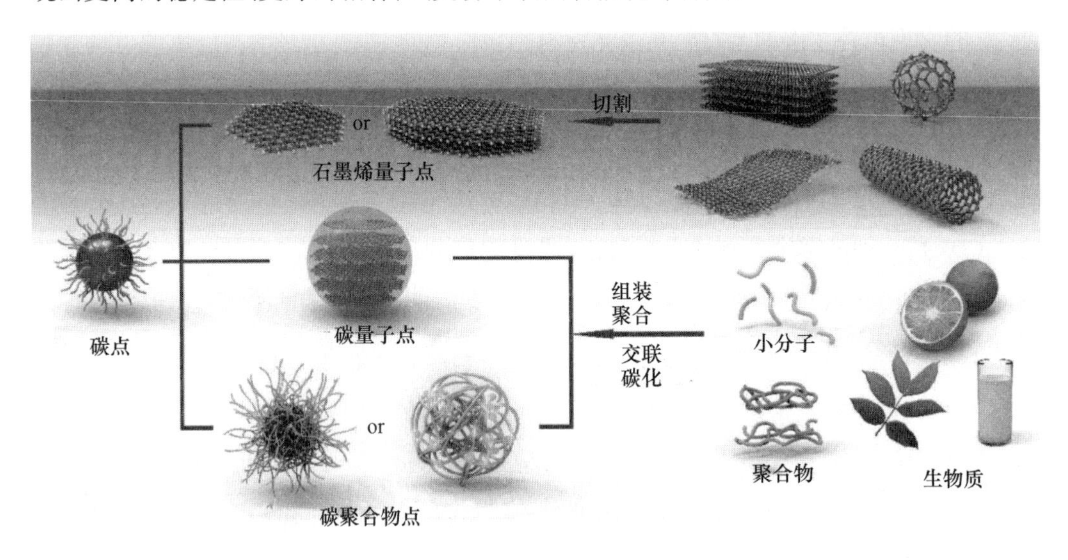

图 1-1 CDs 的分类[26]

1.1.2 碳点的水溶性

对于 CDs 来说,分散性是开展各项研究和应用的基本前提[33]。目前,大部分的 CDs 在水中都具有很好的分散能力,因为它们的表面存在大量含氧官能团[33]。如果想获得疏水 CDs,可以在适当的反应条件下选取合适的疏水前驱体制备得到,或通过疏水分子修饰亲水性 CDs 获得[34]。此外,研究发现通过环境刺激可以实现 CDs 润湿性的可逆控制,这将极大地扩展 CDs 的功能的应用领域[33,35-37]。

1.2 碳点的碳源选择

自 2004 年首次发现 CDs 以来,已有数百种材料被报道用于制备 CDs,这些材料从化学物质到天然产物[38]。起初，制备 CDs 的原料仅仅局限于碳质化合物，如碳粉、蜡烛、石墨和 C_{60}，然而这些原料制备 CDs 的量子产率极低[38]。2006 年，Sun 用有机聚合物（乙二醇 PEG_{1500N}）修饰 CDs，制备出表面附着有机基团的钝化 CDs，在 400 nm 激发时量子产率高达 10%，这表明 CDs 的荧光中心可能来自表面能阱，在表面钝化稳定后产生荧光发射[39]。从那时起，氮掺杂被引入到 CDs 的制备中。例如，以柠檬酸铵、葡萄糖和羟基胺、柠檬酸为碳源，聚乙二胺、乙二胺为氮源制备 CDs[38]。

2012 年，Liu 等人以草为原料，通过水热处理制备了 CDs 并用于检测 Cu^{2+}[40]，自此掀起了从天然物质制备 CDs 的热潮。由于含有氮元素的天然产物种类繁多，橙汁、咖啡渣、家蚕丝、绿茶、鸡蛋、蛋壳膜、豆浆、面粉、香蕉、甜红辣椒、蜂蜜、香菜叶、大蒜、香柠檬、芦荟、玫瑰花和酸橙汽水等原料很容易制备出具有优良荧光性能和较高量子产率的 CDs[38]。

苯二胺 $C_6H_4(NH_2)_2$ 具有两个高活性氨基，这使得苯二胺在高温、低温甚至室温下可以通过氧化/聚合产生各种氧化产物，这些氧化产物进一步交联、碳化或聚合得到 CDs[6,41-44]。苯二胺有三个同分异构体，以不同位置的氨基取代基为碳源制备碳点的 CDs 可以表现出不同的性质，如荧光红移等。这一现象为多功能 CDs 的发展提供了可能。2015 年，Lin 等人使用 PD 的三种异构体（对苯二胺，p-PD；邻苯二胺，o-PD；间苯二胺，m-PD）作为前体，制备出了三种不同荧光性能的 CDs，三种 CDs 通过单个光激发展现出了红、绿、蓝三种荧光发射。这次报道是首次系统地制备多色荧光材料，并对苯二胺同分异构体的荧光机理进行了研究[45]。自此，苯二胺及其衍生物制备 CDs 成了 CDs 制备研究领域的热门方向之一。事实上，以苯二胺为初始碳源制备的 CDs 通常表现出不同于其他碳源的独特光学特性，包括：长波长发射和多色发射。因此，苯二胺衍生的 CDs 已广泛应用于传感、成像等领域[12,46,47]。

1.3 碳点的制备方法

为了获得价格低廉、性能优异的高质量 CDs，多种制备方法被开发。通常这些方法可以根据所选前驱体的尺寸和结构分为两大类，即：自上而下法和自下而上法[8]。通过优化制备条件（例如：制备时间、温度和电压）可以制备出特定性能的 CDs。同时，产物的进一步纯化是获得性能均一 CDs 的重要方式。这些纯化方法种类繁多，从简单处理（例如：离心、透析和过滤）到精确而复杂的技术（例如：电泳、硅胶色谱柱和高效液相色谱等）都可以被利用。此外，制备过程中或制备后在 CDs 表面进行的功能化修饰对 CDs

性能和应用也具有重要意义。尽管如此,如何通过有效的制备方法来精准获得所需性能的 CDs 仍然是一个值得探讨的问题。

1.3.1 自上而下法

自上而下法是将大尺寸碳材料通过物理或化学方法切割得到 CDs 的一类方法总称。自上而下法包括:激光烧蚀法、电化学氧化法、化学氧化法以及超声法等[8]。

1.3.1.1 激光烧蚀法

激光烧蚀已被广泛用于 CDs 的制备,通过激光烧蚀法可以实现对 CDs 的形态调控,并可以改善 CDs 的荧光性能[48,49]。例如:Hu 等人发现调整激光脉冲宽度会对 CDs 的成核和生长产生影响,从而实现了 CDs 大小的控制[50]。同时,他们还提出与短脉冲宽度激光器相比,长脉冲宽度激光器更适合调控纳米材料的尺寸和形貌。Sun 的团队在水蒸气存在的环境中通过激光烧蚀以石墨为碳源制备了 CDs,该 CDs 在溶液和固态时均表现出极强的荧光性能,并有望超越硅基纳米材料[11]。2017 年,Xu 等人在氨基甲苯中采用飞秒激光烧蚀石墨粉制备出了 CDs[51]。在单个激发波长照射下,所制备的 CDs 发射可覆盖几乎整个可见光区域。此外,基于表面基团的作用,这些 CDs 可以用于 pH 传感。

1.3.1.2 电化学氧化法

电化学氧化法制备的 CDs 具有纯度高、成本低、产率高、易于调控尺寸以及重现性好等优点,因此它是获得 CDs 的最常用方法之一[52]。2007 年,Ding 课题组首次采用电化学氧化法在含有高氯酸四丁铵的乙腈溶液中以多壁碳纳米管为碳源制备了 CDs[53]。Liu 等人利用石墨电极在碱性醇中的电化学氧化制得具有高结晶度的 CDs,实验证明施加不同电势和改变 pH 环境对 CDs 的产生具有重要影响[54]。另一方面,一些研究表明可以通过改变电解质的方式获得具有特殊荧光特性的 CDs。例如:Li 等人以氢氧化钠/乙醇作为电解质,通过电化学氧化法制备出具有不同尺寸和不同荧光发射性能的 CDs 的混合物,并通过柱色谱法进行分离[46]。

1.3.1.3 化学氧化法

化学氧化法是指在氧化剂(例如:HNO_3、H_2SO_4 等)的参与下,碳源之间通过氧化剂的诱导发生氧化、聚合制备 CDs 的一种方法。它是用于大规模生产 CDs 有效且方便的方法,并且不需要复杂的设备[55]。同时,氧化反应还能为 CDs 表面引入多种含氧官能团,并赋予 CDs 良好的亲水性和可调的荧光发射性[38]。Pang 等人通过 HNO_3 氧化碳纤维实现了 CDs 的制备[56]。研究发现,通过调节碳源与 HNO_3 的反应时间、温度和浓度,可以对 CDs 表面氧化程度和尺寸大小进行调控,从而获得一系列具有不同荧光颜色的 CDs。Liu 等人还使用磷酸作为氧化剂,在低温下制备出了蓝色和黄色的磷酸盐官能化 CDs[57]。同时,具有高度稳定性的两种磷酸盐官能化 CDs 在生物成像和生物分析领域表现出了巨大的应用前景。Qiao 等人使用三种不同类型的活性炭作为碳源,以 HNO_3 为氧化剂,通过端氨基化合物进行钝化,制备出了三种分散良好的 CDs,并实现了大规模生产[55]。

1.3.1.4 超声法

在液体环境中,超声波会诱导流体动力学剪切力的产生,将宏观碳材料切割成纳米级 CDs[58]。Kang 等人采用一步法,通过过氧化氢辅助超声处理活性炭制备出了水溶性 CDs。制备的 CDs 溶液非常稳定,放置 6 个月没有纳米颗粒沉淀,同时还发现该 CDs 具有上转换荧光特性[59]。此外,Park 等人通过超声处理食物残渣实现了 CDs 的大规模制备,由于制备的 CDs 表面包含了大量的含氧官能团,因此在水中展现了极高的溶解度[60]。Zhuo 等人以石墨烯为碳源,通过浓硫酸和浓硝酸的氧化处理,利用超声法制备了 CDs[61]。所制备的 CDs 表现出不依赖激发波长的下转换和上转换荧光发射行为,并且设计成为光催化剂用于降解亚甲基蓝。

1.3.2 自下而上法

随着对 CDs 研究的深入,研究人员受 CDs 结构的启发,开始利用小分子之间的碳化偶联来制备 CDs,这种方法称为自下而上法。自下而上法是通过有机小分子的热解或碳化制备 CDs 的一种方法。与自上而下法相比,自下而上法具有量子产率高、制备效率高以及大尺寸副产物较少等优点。下面对几种常见的方法进行简单介绍,包括:微波法、热分解法和水/溶剂热法等[8]。

1.3.2.1 微波法

作为电磁波的一种,微波可以瞬间提供密集的能量来破坏碳材料的化学键[62]。因此,微波法可以有效地缩短制备 CDs 的反应时间,使 CDs 的制备变得更加方便和快捷。同时,由于微波的原位和瞬态加热特性,它被认为是制备 CDs 的一种高效省时的方法,会提高产品的产量和质量[63]。2009 年,Zhu 等人首次提出将聚乙二醇和单糖(葡萄糖、果糖)的水溶液在微波炉中加热几分钟即可获得具有稳定的明亮发光以及出色的水分散性的 CDs[64]。Tang 等人结合水热法和微波法的优势,通过微波辅助水热法制备了尺寸可控的葡萄糖衍生 CDs,这也开启了 CDs 制备的新思路[65]。此外,研究发现,当 CDs 涂覆在蓝色 LED 上时,CDs 能够将蓝光转换为白光。2014 年,Zhang 等人以变性蛋白质为碳源微波法制备出了 CDs,并对 CDs 的形成机理进行了研究,最后将其应用于温度、pH 和金属离子检测中[66]。2019 年,Wang 等人以邻苯二甲酸和哌嗪为前驱体,通过微波法制备具有强固态荧光的 CDs,并将其用于快速指纹检测和高质量白光发光二极管的制备中[67]。

1.3.2.2 热分解法

在 CDs 出现前,热分解法已在制备半导体材料和磁性纳米材料中有所使用。进一步的研究表明,热分解法也可以应用于 CDs 的制备中。简单来说,外部提供的热量可以导致有机物的脱水和碳化,最终这些脱水和碳化产物将转变为 CDs。该方法具有操作简单、无溶剂参与、反应时间较短、成本低廉以及可大规模生产等优点[68,69]。Ma 等人通过沙浴加热将乙二胺四乙酸直接碳化成具有石墨烯结构的杂原子掺杂 CDs,制备的 CDs 具有强荧光性、良好的 pH 稳定性和低细胞毒性,适用于生物标记和光电应用,并且该方法可以扩展到更广泛的前驱体范围,为不同类型的杂原子掺杂 CDs 提供了新途径[70]。Martindale 的团队采用热分解法,在 180 ℃下通过直接热解柠檬酸制得产率高达 45% 的 CDs,研究发现 CDs 可作为光敏剂用于光催化产氢,在超过 455 nm 的可见光区域仍具有活性,并在完整的太阳光谱辐射下保持至少 24 小时的完全光催化活性[71]。

1.3.2.3 水/溶剂热法

水/溶剂热法是一种低成本的制备途径。该方法所用碳源广泛,从糖类、有机酸到生活中的果汁或果皮都可以作为碳源。因此,水/溶剂热法也是目前制备 CDs 最常用的方法之一。通常,水/溶剂热法是将前驱体的溶液密封在反应釜中,通过高温高压反应制得 CDs。2013 年,Zhu 等人通过水热法制备了一种具有高量子产率的荧光 CDs,产生的 CDs 已被用于印刷油墨和 Fe^{3+} 生物传感器的构建[12]。Jiang 等人选取对苯二胺、邻苯二胺和间苯二胺分别作为碳源,以乙醇为溶剂,制备出了红色、绿色和蓝色荧光发射 CDs[45]。他们认为这些 CDs 的荧光发射变化与它们的粒径和氮含量有关。同时,研究发现可以通过适当的比例混合两种或三种 CDs 来获得柔性的全色发光膜。另外,这些 CDs 还显示出低细胞毒性和出色的细胞成像能力。

1.4 碳点的光学性质

1.4.1 紫外吸收特性

采用不同碳源作为前驱体或通过不同的制备方法得到的 CDs 会表现出不同的吸收行为。然而,它们通常在紫外范围内表现出很强的吸收,并伴随着一条延伸至可见光范围内的尾巴,这归因于 C = C 键的 $\pi-\pi^*$ 跃迁或 C = O/C = N 键的 $n-\pi^*$ 跃迁[12,72-75]。此外,某些发出红光或近红外光的 CDs,通常其 sp^2 域的 π 共轭电子和所连接表面基团或聚合物链中的 π 共轭电子会引起长波段吸收[76,77]。因此,CDs 的吸收特性主要受表面基团的类型和含量、π 共轭电子以及碳核中氧和氮的基团影响[26]。

1.4.2 下转换荧光特性

下转换荧光,也就是激发波长小于发射波长的传统荧光发射,本书中所提及的光致发光在无特殊说明下均为下转换荧光,简称荧光。荧光性质是 CDs 在基础研究和应用中最吸引人的属性之一。与其他荧光材料(例如:镉/铅量子点、稀土纳米材料和有机染料等)相比,CDs 具有更好的光稳定性、更高的量子产率、更低的毒性、低廉的成本和出色的生物相容性等优点,这使其在多个领域具有广泛的应用[8]。

从基础和应用的角度来看,下转换荧光发射是 CDs 最吸引人的特性之一。2015 年,Lin 等人报道了对苯二胺、邻苯二胺和间苯二胺的乙醇溶液通过溶剂热法获得红色、绿色和蓝色发射 CDs[45]。这个发现推动了 CDs 研究进入了一个新的时代。表 1-1 列出了以苯二胺为前驱体制备 CDs 的方法、λ_{em} 波长、荧光量子产率和尺寸等方面的最新研究成果。从表 1-1 可以看出,随着使用不同的苯二胺同分异构体,CDs 的荧光发射逐渐发生红移,这可能与 CDs 的结构有关[13]。

表 1-1 苯二胺衍生碳点的制备方法、λ_{em} 波长、荧光量子产率和尺寸

碳源	制备方法	λ_{em} 波长 /nm	荧光量子产率/%	尺寸/nm	参考文献
间苯二胺	溶剂热法	488	31.58	5 ~ 6	[78]
间苯二胺、磺酰胺	水热法	520	78.6	2.7 ± 0.9	[79]

续表

碳源	制备方法	λ_{em}波长 /nm	荧光量子产率/%	尺寸/nm	参考文献
间苯二胺/邻苯二胺/对苯二胺	微波法	520/ 573/ 402	16.72/ 38.5/ —	—/8 ~ 11/ —	[80]
邻苯二胺、柠檬酸	微波法	565	—	5.0 ±0.4	[81]
邻苯二胺、柠檬酸/间苯二胺、柠檬酸/对苯二胺、柠檬酸	太阳照射	443/ 509/ 592	81/ 47/ 1.2	2.03/3.67 /4.38	[82]
邻苯二胺、对苯二酚	氧化法	545/ 555	22.4/ 16.9	20.9/ 20.7	[6]
邻苯二胺	水热法	567	2.0	4.0 ±0.3	[83]
邻苯二胺、硫脲	溶剂热法	556	26	3.75	[84]
邻苯二胺、磷酸	水热法	568/ 622	34/ 15	2.91/ 2.99	[85]
邻苯二胺、色氨酸	水热法	560	20.63	10.8	[86]
邻苯二胺、3 - 氨基苯基硼酸	水热法	645	8.56	4.09	[87]
邻苯二胺、双氰胺	水热法	630 和 680（双发射）	30.2	5.71	[88]
邻苯二胺、甲酰胺	水热法	556	20	3.32	[89]
邻苯二胺、4 - 1,2 - 苯二胺、对苯醌	氧化法	530/ 555/ 600	—/—/ —	5.3/ 5.4/ 5.2	[41]
邻苯二胺、多巴胺	氧化法	572	—	3	[90]
对苯二胺	水热法	618	3.2	5	[91]
对苯二胺、N - [3 - (三甲氧基硅基)丙基]乙二胺	水热法	498	13.83	3.8	[92]
对苯二胺	溶剂热法	615	13.58	2.0 ~ 6.5	[93]

续表

碳源	制备方法	λ_{em}波长 /nm	荧光量子产率/%	尺寸/nm	参考文献
对苯二胺、二乙烯三胺五乙酸	水热法	610	5.76	2 ~ 5	[94]
对苯二胺、4 - 甲酸基苯硼酸	微波法	606	5.5	10	[95]
对苯二胺、草酸	溶剂热法	609	—	2.66	[96]
对苯二胺、半胱胺盐酸盐	溶剂热法	550	—	5	[97]
对苯二胺、磷酸、$Mn(OAC)_2$	微波法	600	30.1	6.47	[98]
对苯二胺、叶酸	水热法	470	21.8	2.0 ± 0.6	[99]
对苯二胺、$FeCl_3$	氧化法	600	7	1.2 ~ 2.8	[100]
对苯二胺、氨基苯甲酸	水热法	653/ 621	18	12	[101]

1.4.2.1 单发射

研究表明,由三种苯二胺同分异构体制备的 CDs 具有不同的荧光发射分布,其中间位取代通常产生发射波长较短的 CDs(蓝色发射)。例如,Lin 等人在 180 ℃的乙醇溶液中加热间苯二胺 12 小时,制备出最大发射波长为 435 nm 的 CDs[45]。Li 等选择叶酸和间苯二胺制备 CDs,在紫外区激发下 CDs 发射出稳定的蓝色荧光,最大发射长度为 429 nm (λ_{ex} = 365 nm),相应的量子产率为 12.6%[102]。蓝光发射限制了 CDs 在生物医学上的进一步应用[103]。因此,长波长发射 CDs 的制备备受关注。

o - PD 衍生 CDs 的荧光发射通常≥500 nm。Lu 等人提出了一种基于黄色发射 CDs (λ_{em} = 564 nm)的传感器用于氟喹诺酮类和组氨酸的检测,此 CDs 由 o - PD 和 4 - 氨基丁酸制备而成,具有良好的光学和生物学特性,包括:化学稳定性高、生物相容性好、细胞毒性低等[104]。2019 年,Li 等报道了以 o - PD 为碳源,使用不同的反应溶剂制备了一系列发射波长约为 540 nm 的 CDs[105]。Sun 等以 o - PD 和正硅酸乙酯为原料,通过一步水热法制备了掺硅的橘黄色荧光发射 CDs,该 CDs 在 580 nm 处有一个明亮的荧光发射峰,同时具有良好的生物相容性和荧光稳定性[106]。此外,掺杂会影响 o－PD 衍生 CDs 的荧光发射。例如,Gao 等人报道了一种以 o－PD 为前驱体的溶剂热制备 CDs 的方法,基于硫掺杂可调节碳点的荧光发射从绿色到红色,这是因为硫的掺杂可以显著增加石墨氮的含量,才会引起荧光发射发生改变[107]。最近,有文献报道 o－PD 衍生 CDs 的荧光发射突破了 600 nm。Zhang 等人通过质子化处理 o－PD 衍生 CDs,获得了具有单光子红色荧光发射(620 nm)和近红外的双光子红色荧光发射(630 和 680 nm)性能的 CDs[108]。实验结果证实,2,3－二氨基苯肼荧光团的质子化改变了 CDs 的分子状态,减小了光子跃迁带隙,触发了单光子和近红外诱导的双光子红色发射。由于 2,3－二氨基苯肼荧光团是 o－PD 氧化产物,他们提出 2,3－二氨基苯肼决定了 CDs 的荧光发射,这意味着从前驱体转化的荧光团产物可以作为具有所需荧光特性的 CDs 的预测因素。

选取对苯二胺为碳源制备的 CDs 通常表现出较长的发射波长,Zhang 等人提出了一种溶剂热法,以对苯二胺和半胱胺盐酸盐为前驱物,制备出了一种在 555 nm 处具有最大荧光发射波长的 CDs[97]。此外,近期的研究表明,用对苯二胺作为前驱体制备的 CDs 发射波长可以超过 600 nm。Xiong 等人使用对苯二胺和尿素作为前驱体,通过水热法制备了具有可调荧光和量子产率达 35% 的 CDs。实验发现,在单波长紫外光照射下,随着制备的 CDs 结构中含氧量增加,荧光发射逐渐发生红移,最大发射值为 625 nm[13]。Liu 等以对苯二胺为碳源,通过调节反应前 HNO_3 的加入量来调控 CDs 的荧光性质,研究发现制备的 CDs 的发射波长可达 630 nm[31]。最近,人们发现可以通过调节前驱体溶液的 pH 值制备多彩 CDs。Jiao 等人证实,随着前驱体溶液 pH 从碱到酸的调节,由对苯二胺制备得到的 CDs 的荧光发射从 515 nm 逐渐变化到 615 nm,他们认为 CDs 中荧光发射的多样化源于其颗粒大小和氮掺杂含量。最后,他们推测对苯二胺上的氨基在酸性条件下更容易被激活,导致深度交联,促进长波长发射[109]。Hua 等发现以对苯二胺为碳源水热法制备 CDs 的过程中加入金属离子会影响制备产物的荧光发射,最后可形成发射波长高达 700 nm 的 CDs[110]。虽然金属离子在 CDs 的形成过程中起着至关重要的作用,但在得到的 CDs 中却不存在金属离子,即 CDs 是不含金属的,金属离子在 CDs 的形成过程中起着类似于“催化剂”的作用。此外,以对苯二胺和镍离子为前驱体制备了 CDs,具有高度的生物相容性,可以实现实时和高分辨率的细胞核成像,以及荷瘤小鼠和斑马鱼的高对比度成像。

1.4.2.2 双发射

为了扩大 CDs 在生物传感器和生物图像中的应用,双发射 CDs 受到了极大关注,这主要是因为双发射 CDs 可以有效地消除环境变化和探针浓度变化的影响。

Song 等人采用水热法,在 200 ℃ 下处理 o - PD 和磷酸的混合物 24 小时制备出 CDs,该 CDs 在 380 nm 处激发时于 440 nm 和 624 nm 处显示出两个独特的荧光发射峰。同时,该 CDs 表现出了对赖氨酸(440 nm)和 pH(624 nm)的双重响应功能[111]。Yuan 等人使用对苯二胺和甲酰胺进行溶剂热制备得到橙/蓝色发射 CDs,他们认为氮氧相关基团是调控 CDs 双发射的原因[112]。Chen 的团队通过调节 o - PD 和赖氨酸的质量比,制备了双发射 CDs,他们发现随着引入赖氨酸质量的增加,o - PD 聚合物链的碳化程度逐渐被抑制,最终 sp^2 共轭域消失,也就是说双发射 CDs 是通过赖氨酸质量变化引起的[113]。Pan 等人开发了一种方法,将没食子酸和 o - PD 在水溶液中碳化,制备出分别在 470 nm 和 570 nm 处具有双发射峰的 CDs,制备的双发射 CDs 可用于比率检测 Fe^{3+},并可通过焦磷酸盐恢复 Fe^{3+} 引起的荧光淬灭,他们进一步利用 CDs - Fe^{3+} - 焦磷酸盐杂化体系检测酸性磷酸酶,也获得了较好的效果[114]。Li 等报道了双发射 CDs 用于检测兽药,在他们的报道中使用山梨醇和 o - PD 作为前体制备 CDs,该 CDs 在 597 nm 和 645 nm处表现为双荧光发射,经过表征未经进一步表面处理的 CDs 的量子产率为 33%[115]。Yang 等人选取 o - PD 为前驱体,$Al(NO_3)_3 \cdot 9H_2O$ 为辅助剂制备 CDs,实验发现制备的 CDs 在 600 nm 和 650 nm 条件下表现出红色双发射[116]。Huang 等人利用微波法处理o - PD 获得了在 320 nm 的激发下于 360 nm 和 530 nm 双发射 CDs[117]。从上述的实例可以看出,双发射 CDs 在整个 CDs 领域是非常热门的一个方向。

1.4.2.3 量子产率

量子产率是衡量光化学反应中光量子利用率的指标。在光化学反应中,特定的波长下进行光化学反应的光子与吸收总光子数之比即为量子产率。量子产率值能够定量地反映出荧光的亮度。在 CDs 领域中,量子产率受碳源、制备方法和钝化的影响。从广义上讲,"自上而下"路线制备的 CDs 与"自下而上"路线相比通常具有较低的量子产率[2,16,26,31]。从 CDs 的形成和荧光机理的角度来看,在切割宏观碳源的过程中会形成更多的缺陷,导致 CDs 的量子产率较低。获得量子产率的方法分为两种,一种是通过仪器直接测得的,称为绝对量子产率;另一种是将已知量子产率的物质作为参考计算求得的,称为相对量子产率。

1.4.2.4 荧光机理

CDs 被发现后不久,研究人员便开始思考 CDs 的发光机理。发光机理的研究对于设计具有可调荧光发射的 CDs 具有重要意义。截至目前,尽管已经报道了许多关于

CDs 荧光机理的研究工作,但是对于 CDs 荧光起源仍然存在一定的争议。由不同成分和方法制备的 CDs 通常包含不同的组成和复杂的结构。也就是说,使用不同的制备方法、前驱体和后处理制备的 CDs 表现出不同的光学性能,这表明 CDs 的形成是一个比预期更复杂的过程。因此,通过比较不同文献中报道的 CDs 性质以建立统一的理论是不恰当的。当前,普遍接受的发光机理主要有三种,包括:表面态理论、量子限域效应以及分子荧光[4]。

(1)表面态理论。

表面态理论是迄今为止被广泛接受的发光机理之一,这里所指的表面态包括表面氧化程度和表面官能团种类、数量。许多研究表明,CDs 的荧光受控于其表面氧化程度。CDs 的发射红移与它们表面的氧化程度密切相关。表面氧化度越高,表面缺陷的数量越多,这些缺陷会捕获电荷,并且由捕获的电荷复合产生的辐射会引起发射红移。Xiong 的团队通过柱色谱法获得了一系列纯化的 CDs(图 1-2),研究发现这些 CDs 的表面氧化程度随着荧光发射的红移而逐渐增加[13]。表面氧化会产生表面缺陷,该缺陷充当电子的捕获中心,从而引起与表面态有关的荧光。另一方面,随着 CDs 表面氧含量的增加,其带隙会减小。也就是说,红移发射源于表面氧化程度的增加[56]。Liu 等人设计了一种在室温下制备不同发射 CDs 的简单且环保的方法。研究发现,具有相同粒径大小和官能团的 CDs 表现出不同程度的表面氧化[4]。随着表面氧化程度的增加,发射波长逐渐红移,这种现象是 CDs 中的 LUMO 和 HOMO 之间的带隙减小所引起的。Zhu 等也证明了 CDs 表面氧化程度增大是造成红移的原因,且氧化后 CDs 的荧光发射性能与硅纳米晶体相同[118]。

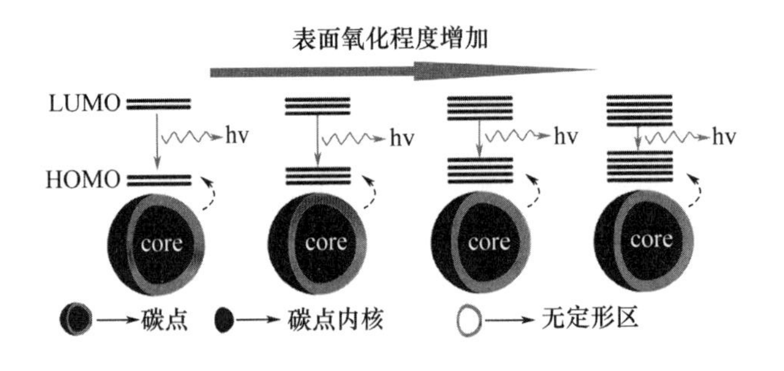

图 1-2　基于表面氧化程度的 CDs 的荧光机理[13]

除了表面氧化程度,一些研究人员将表面态变化与表面官能团组成相联系,不同的表面官能团可以在 CDs 中引入不同的荧光团或能级进而影响 CDs 的发光性能。Sun 等人证明通过调节 CDs 表面的官能团,CDs 的发射波长可以显著变化[119]。此外,他们还发现 CDs 的发射中心均匀地分布在一些特殊的边缘态,这些边缘态由几个碳原子和碳骨架边缘上的官能团(例如:羰基)组成。因此,CDs 表面的官能团最终可以通过影响发射中心来调节 CDs 的荧光。此外,Liang 的研究团队报道了一种多色荧光发射 CDs,其荧光可调范围从深蓝色到红色甚至白色(图 1 - 3)[120]。由于制备的 CDs 具有相似的氧含量,因此,它们的荧光发射取决于表面官能团而不是表面氧化程度。对于具有各种表面态的 CDs,除了 HOMO - 2(π)能级之外,诸如 C = O 和 C = N 的官能团可以引入两个新的能级(HOMO - 1 和 HOMO)并进一步产生新的电子,引发从 HOMO - 1 和 HOMO 能级到 LUMO(π *)的跃迁。因此,当处于氮相关缺陷状态的电子返回到 HOMO 时,荧光发射可以转移到长波长范围。

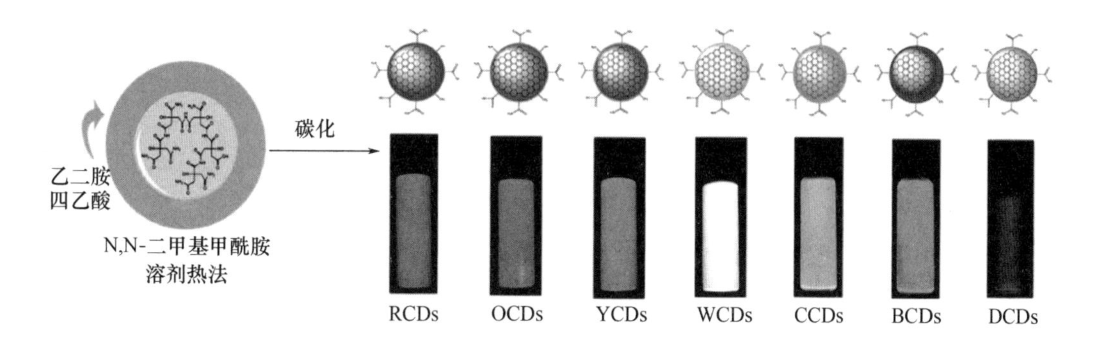

图 1 - 3　制备具有不同官能团碳点示意图[120]

(2)量子限域效应。

当半导体晶体为纳米级尺寸时,晶体边界会显著影响电子分布,导致半导体晶体显示出某些性质,例如:带隙和能量弛豫动力学尺寸依赖性[121]。在许多半导体材料中已经研究了这种现象,并将其称为量子限域效应。CDs 作为一种新的零维量子限域体系,由于石墨烯的维数和元素组成的影响,也会受到量子限域效应调控而使其具有某些独特的性质[122]。

Choi 等制备了一系列具有不同尺寸和不同形态的 CDs,系统地研究了 CDs 的尺寸与荧光发射之间的关系(图 1 - 4)[123]。研究发现,随着 CDs 尺寸的增加,CDs 的吸收峰能量减小,这是由量子限域效应引起的。此外,他们还提出了 CDs 荧光发射的红移是 CDs 的尺寸增加所致。Jiang 等人制备了三种不同荧光发射 CDs,在紫外光激发下分别具有红色、绿色和蓝色荧光[45]。三种 CDs 具有相似的化学组成,但粒径大小分布却不同。因此,基于量子限域效应,他们提出了 CDs 的荧光发射波长改变与 CDs 尺寸密切相关。此外,Lee 等人发现,即使通过氢等离子体除去 CDs 的表面氧元素,CDs 的荧光也不会改变。最后,他们提出 CDs 的荧光是基于量子限域效应的结论[46]。

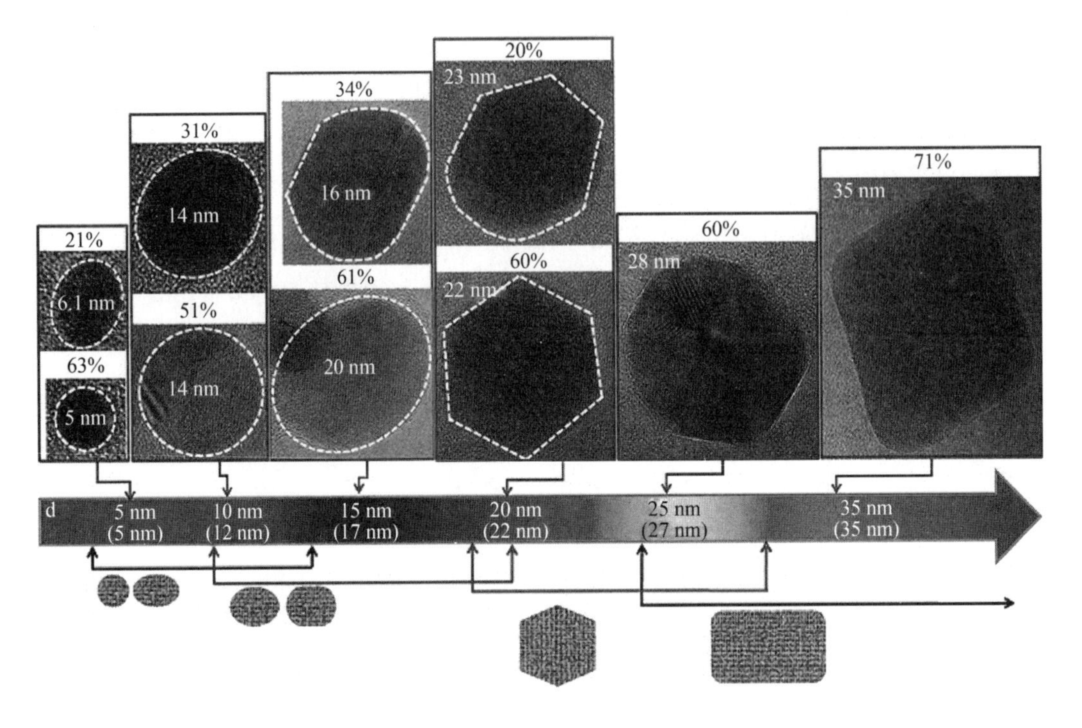

图 1-4 CDs 尺寸和荧光发射关系示意图[123]

(3)表面态和量子限域效应协同作用。

目前,有报道证明表面态和量子限域效应会共同决定 CDs 的荧光发射[124]。Pang 课题组提出 CDs 的带隙由其表面性质和粒径大小决定[56]。通过改变 CDs 表面氧化程度或粒径大小,可以制备一系列具有不同荧光的 CDs。结果表明,CDs 粒径的增加或表面氧化程度的增加都会引起 CDs 的荧光发射红移。Huang 课题组也提出 CDs 的荧光受量子限域效应和表面态的协同调控[4]。他们通过水热法制备出了一系列的全色发射 CDs,并通过硅胶色谱柱进一步分离。研究发现,CDs 的荧光性质既可以根据量子限域效应通过调节晶体碳核的大小来粗调,也可以通过表面态调控来改变 CDs 表面官能团来细调。

(4)分子荧光。

研究表明,在自下而上的化学制备过程中,荧光小分子的形成(即分子荧光)也会影响 CDs 的发射[125]。Yang 课题组证明使用柠檬酸和乙二胺制备的 CDs 的强荧光发射来自蓝色荧光分子,同时还证明了 CDs 是由荧光分子、聚合物和碳核组成的混合物(图 1-5)[126]。此外,Rogach 课题组通过制备 3 种基于柠檬酸的 CDs,阐明了不同分子荧光对 CDs 发射的贡献,证明了不但在溶液中游离的荧光分子团对 CDs 的光学性质有很大影响,而且附着在 CDs 上的荧光分子对于 CDs 的发光性也具有举足轻重的作用[127]。

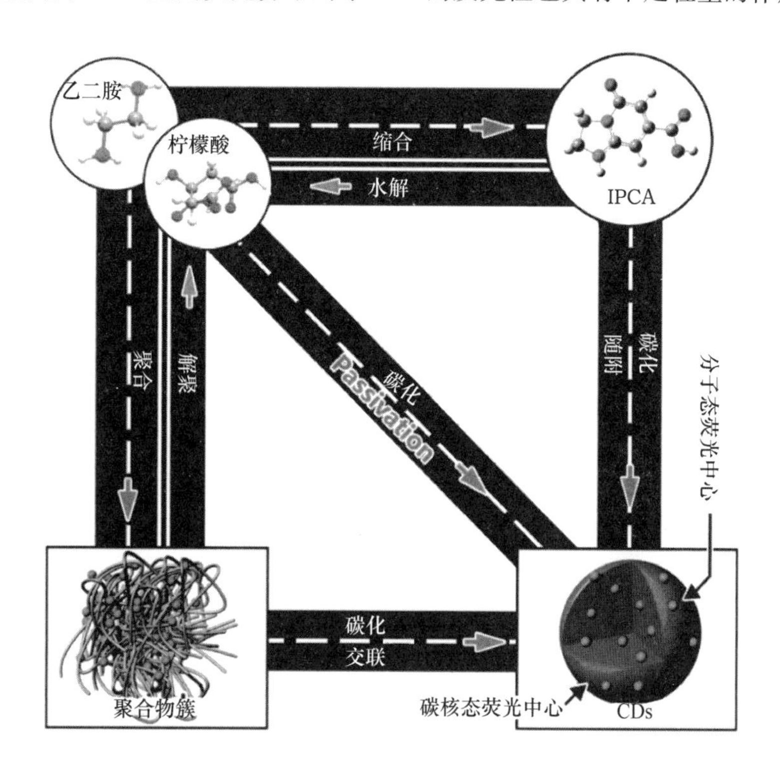

图 1-5 柠檬酸和乙二胺制备 CDs 体系中不同产品之间关系示意图[126]

最近,Baker 团队通过去除荧光分子团证明作为 CDs 合成副产物的荧光小分子是荧光发射的主要来源[125]。同样地,Ferrante 课题组证明 CDs 的荧光来自自由分子[128]。分散在溶液中的小荧光分子引起了 320 ~ 450 nm 范围内激发的 CDs 发射。而在 480 nm 以上激发时的发射主要由发射性能微弱的碳核决定。同时,尽管实际上存在碳核,但是荧光主要来自溶液中游离的有机小分子。

1.4.3 上转换荧光特性

研究表明,除了常规的荧光发射性能外,某些 CDs 还具有上转换荧光(UCPL)发射的特征。在下转换荧光中,发射波长的能量是低于激发波长的。但是,UCPL 发射与下转换荧光正好相反。也就是说,发射波长的能量是大于激发波长的[49]。2007 年,Cao 等人通过激光烧蚀制备了在近红外(800 nm)处显示出具有双光子激发的强发光 CDs,从而证明了 CDs 具有 UCPL 特性[129]。此后,其他课题组也陆续报道了 CDs 的 UCPL 发射性能[130-132]。Li 等以葡萄糖为碳源,通过一步碱/酸辅助超声法直接制备了 CDs[132]。

研究发现,所制备的 CDs 除了具有近红外荧光发射外,还具有 UCPL 特性。图 1-6 显示了此 CDs 在长波长光激发下(700～1 000 nm)的 UCPL 发射(300～800 nm)。

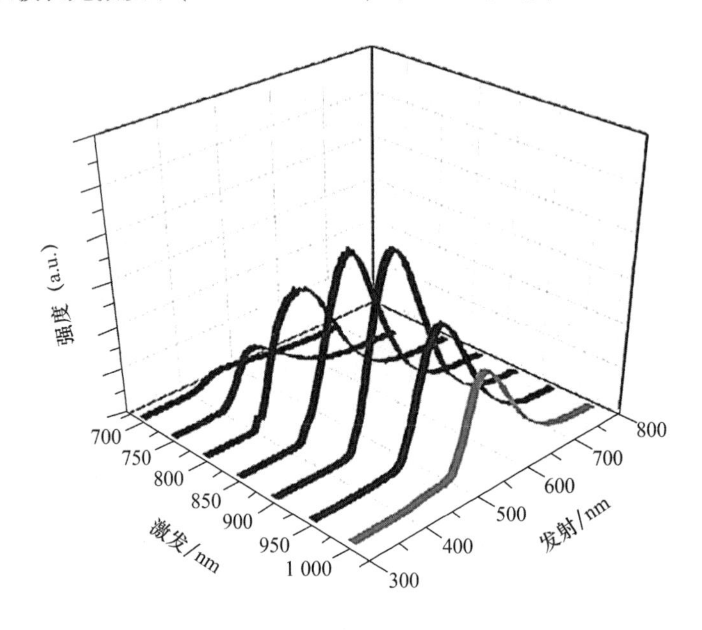

图 1-6 CDs 的上转换荧光光谱[132]

自 CDs 的 UCPL 特性被发现以来,科研人员便开始研究其发射机理。然而,至今人们对于 CDs 具有这种特殊的荧光特性的机理尚未完全清楚。2007 年,Cao 课题组通过观察到的 CDs 的发光强度对激发激光功率的依赖性,首次提出 CDs 的 UCPL 发射机理来自一种双光子机制[129]。后来,Lin 和 Gong 等人进一步通过实验论证了这一假设[45,133]。然而,Shen 等人认为双光子激发不足以解释 CDs 的 UCPL 发射特性[134]。研究发现,随着激发波长增加,CDs 的 UCPL 发射波长也随之改变。但是,激发光能量和 UCPL 能量之间的差值却恒定不变。基于此,他们建立了一个 UCPL 能级结构模型,并提出 CDs 中存在 π 和 δ 两个基态轨道。当光子激发 π 轨道上的电子时,电子会跃迁到高能量轨道。之后,电子再跃迁回低能级轨道。如果跃迁回 δ 轨道,就会发生 UCPL 现象。此外,δ 轨道上的电子也会受到激发发生跃迁,但是它只能够产生常规的荧光发射,这便可以解释 UCPL 中激发波长和发射波长的能量差为一个常数的现象了。

1.4.4 磷光特性

室温磷光(RTP)是 CDs 吸引人的光学性质之一,它是通过两个关键过程产生的:①从最低激发单重态(S_1)到三重态(T_n)的系间窜越(ISC);②从最低激发三重态(T_1)到基态(S_0)的辐射跃迁[135]。因此,有效的ISC过程和抑制非辐射衰变对于机体产生RTP至关重要。目前,已经证实C=O和C=N有助于RTP的产生,因为它们强大的自旋-轨道耦合能引起低单重态-三重态分裂能量[136,137]。同时,氮、磷和卤素等杂原子的掺杂可以促进C=O和C=N的n-π*跃迁,有利于ISC过程。此外,为了获得RTP,也有使用聚乙烯醇和滤纸等基质固定荧光材料,从而稳定产生的T_1,防止CDs三重态的淬灭[23,137,138]。Lin课题组报道了一种非常简便的方法来制备在自然环境下表现出多色荧光并具有较长寿命的RTP-CDs(图1-7)[24]。研究发现,CDs粉末的荧光和RTP颜色分别从蓝色变为绿色,从青色变为黄色。进一步的研究表明,多色发射源自于CDs中多个发射中心的存在,并且较高的制备温度对于获得RTP至关重要。最后,鉴于此RTP-CDs独特的光学性能,将其应用于防伪中。虽然对于RTP-CDs的制备已经取得了很大的进展,并在多模式防伪和生物成像中得到了应用[21,22,24],然而基于CDs的RTP材料的制备和应用尚处于起步阶段,需要通过合理的设计结构、后处理以及反应条件的调控,进一步提高RTP-CDs的量子产率、荧光寿命和稳定性。

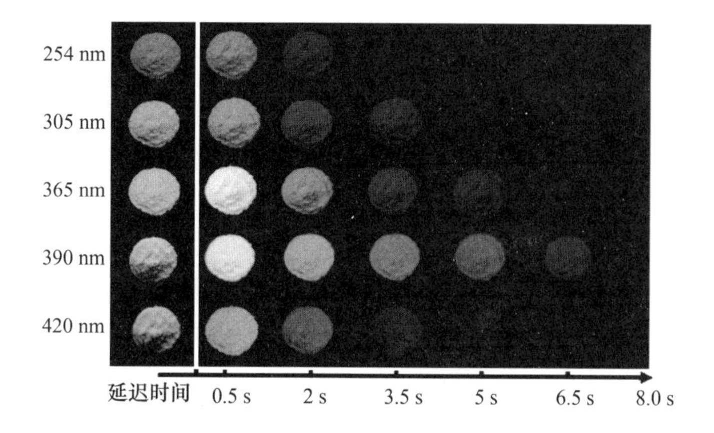

图1-7 不同激发光下CDs的荧光和室温磷光[24]

1.5 碳点的表面态/结构调控

1.5.1 化学掺杂

随着对CDs的进一步研究,人们开始追求更科学、绿色和表面态/结构可控的制备CDs的方法。在此背景下,掺杂作为一种简单的方式进入大家的视野。根据引入CDs

中的杂原子种类的数量,可以分为单掺杂和共掺杂(图 1-8)[139]。研究发现,通过向 CDs 中引入杂原子(例如:氮、硼、硫和磷等)可以有效调整 CDs 电子结构,进而影响 CDs 性质,这被认为是调节 CDs 固有性质的有效方法。因此,在过去一段时间已经有许多的研究工作致力于开发新掺杂剂和掺杂工艺用于调控 CDs 的性质。

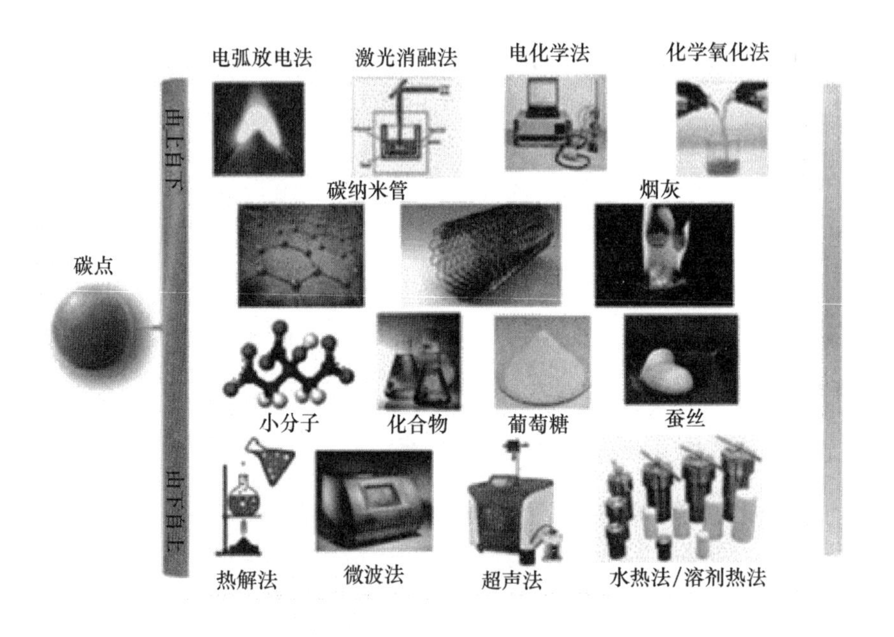

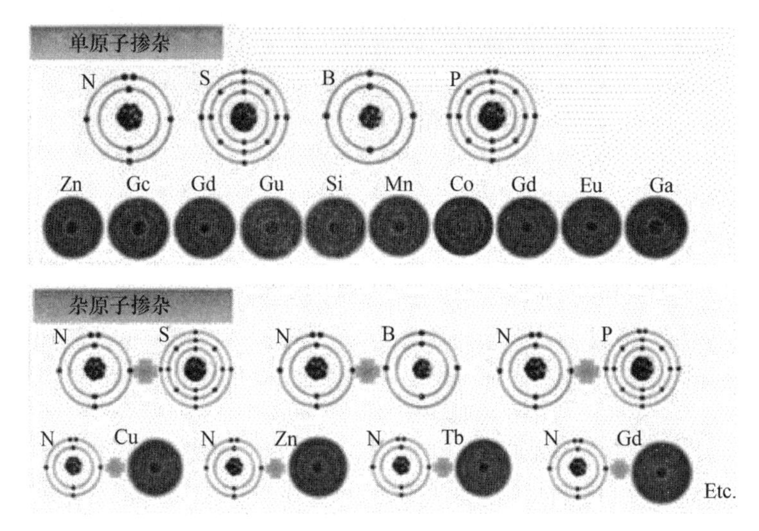

(a)

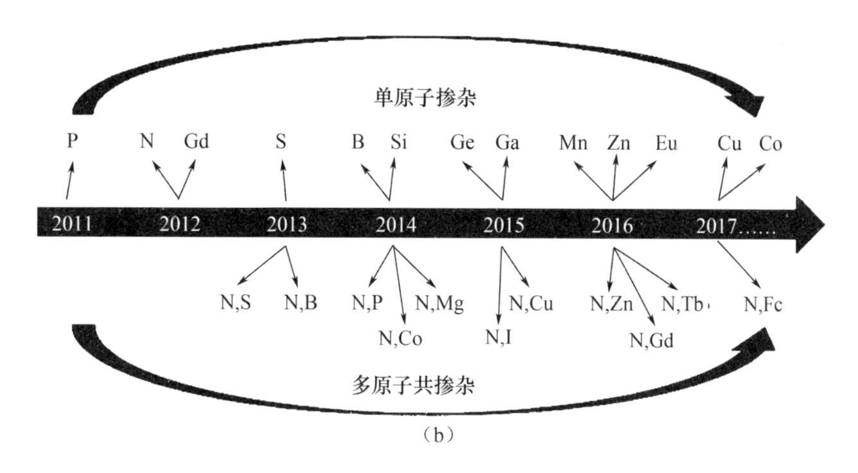

图1-8 制备杂原子掺杂CDs的主要方法[139]

1.5.1.1 氮掺杂

氮是一种典型的掺杂剂,因为它具有5个价电子并且原子大小与碳相当,所以氮掺杂是提高CDs荧光性能最广泛的使用方法。氮原子能够将电子导入CDs中并改变其内部电子环境,从而有效地改善它们的荧光特性。通过这种方法获得的氮掺杂CDs(N-CDs)在生物医学等领域显示出了优异的性能[140]。此外,为了将氮结合到碳框架中,科研人员已经开发了多种方法。

2016年,Li等人报道了一种基于N-CDs的磷光纳米材料,在280 nm激发下展现了1.06 s的超长磷光寿命,磷光量子产率为7%。研究发现,N-CDs表面的C=N基团可以创建新的能级结构,并且首次提出N-CDs的磷光来自C=N基团。同时,这些基于CDs的磷光材料在白光LED和数据安全方面显示出了巨大潜力[141]。Zhang等人通过改变掺杂氮的含量制备出了荧光可调的N-CDs,所获得的CDs具有较高的量子产率和稳定性,无须进一步的功能化就可以直接用于小鼠腹膜巨噬细胞的成像[142]。2016年,Peng等人通过调控前驱体的方式制备了4种不同发光N-CDs。研究发现,氮掺杂对这些CDs的光学性质、pH响应性和光电性质都有影响。最后,制备的N-CDs作为敏化剂,以0.53%的最高光伏转化效率用于构建纳米TiO_2太阳能电池[143]。2019年,Liu等人以梨汁和乙二胺为前驱体,利用微波法制备了N-CDs[144]。他们使用密度泛函理论、拉曼光谱等表征方法系统地研究了氮掺杂诱导的CDs荧光增强机理。同时,基于CA125适配体在N-CDs表面通过竞争性吸附和解吸所诱导的荧光恢复,提出了一种高灵敏度、高选择性的无标记荧光检测法,用于CA125的分析。此外,所提出的荧光检测法还成功用于检测健康人和卵巢癌患者血清样品中的CA125。

1.5.1.2 硫掺杂

近年来,硫掺杂 CDs(S - CDs)得到了越来越多的关注,因为硫原子的加入可以改变 CDs 的电子结构,为光激发电子的捕获提供能量。另外,S - CDs 可以保留未掺杂 CDs 的几乎所有优点。同时,因为其较大的整体 Stokes 位移,所以可以进一步避免自淬灭。目前,S - CDs 的制备方法有许多,包括:水/溶剂热法、热解法、微波法和火焰合成法[145 - 149]。

2020 年,Gao 等人通过硫掺杂制备出了一种荧光可调的 S - CDs。详细的表征和比较表明硫掺杂可以显著增加石墨氮的含量,从而导致 CDs 荧光从绿色变为黄色。由于具有多色发射,所制备的 CDs 可应用于荧光膜、LED、生物成像探针和金属离子检测等领域[107]。Hua 等人通过使用聚(4 - 苯乙烯磺酸 - 共聚 - 马来酸)作为碳源和硫源,通过水热法制备了 S - CDs[150]。基于诺氟沙星/环丙沙星与 CDs 之间强烈的氢键相互作用和电荷转移,将其作为探针用于诺氟沙星/环丙沙星的定量分析。

1.5.1.3 硼掺杂

硼元素在元素周期表中与碳元素相邻。与硫和氮一样,硼也可以共价掺杂到 CDs 中并充当有效的掺杂剂。普遍认为,硼掺杂是改变 CDs 电子结构和光学性质的有效方法之一。因此,硼经常作为掺杂剂用于制备碳基纳米材料。

2014 年,Shan 等人报道了通过溶剂热法来制备 B - CDs,同时利用 B - CDs 与过氧化氢之间发生的电荷转移所引起的荧光淬灭,将 B - CDs 用于定量检测过氧化氢和葡萄糖的含量[151]。在 2015 年,Bourlinos 课题组通过微波法制备了具有准球形和超小尺寸的 B - CDs。与未掺杂的 CDs 相比,B - CDs 表现出了显著的光信号增强[152]。Shen 等人通过水热法制备了产率高达 50% 的 B - CDs,所获得的 B - CDs 在固态和水溶液状态下均显示出强荧光,并具有优异的水溶性。因此,B - CDs 可用作细胞显像剂和发光器件转换材料[153]。2021 年,Xiong 课题组以硼酸为硼源制备了产率高达 75% 的 B - CDs。研究发现,B - CDs 的固态荧光会受硼原子浓度淬灭效应调节。之后,他们将 B - CDs 荧光可视化试剂用于潜指纹显现,显示出了很好的效果[154]。

1.5.1.4 磷掺杂

与氮相反,磷原子的尺寸大于碳原子。因此,磷可以在碳簇中形成取代缺陷,进而改变 CDs 的电子结构和光学性质。目前,磷掺杂 CDs(P - CDs)的制备方法包括:微波法、溶剂热法等[155,156]。

在 2011 年,Chandra 等人报道了在 100 W 下通过微波处理蔗糖的磷酸溶液制备得到高荧光 P－CDs。P－CDs 的大小在 3～10 nm 之间,在紫外线激发下 P－CDs 可以发出亮绿色的荧光。并且,P－CDs 显示了极低的细胞毒性,可以用作生物成像和药物运输的材料[155]。2014 年,Zhou 等人通过溶剂热法制备 P－CDs。研究发现,磷原子的掺杂可以提高发射效率并调节 CDs 的发射带。同时,P－CDs 具有出色的生物标记能力和低细胞毒性,也具有作为光学材料和催化装置材料的潜力[156]。

1.5.1.5 金属原子掺杂

除上述非金属元素外,还有一些金属元素,例如:锌(Zn)、锗(Ge)、钆(Gd)、铜(Cu)、硅(Si)、锰(Mn)、钴(Co)、铕(Eu)和镓(Ga)等,也可以作为掺杂剂用于调控 CDs 的电化学、生物学和光学性质[139]。

2016 年,Xu 等人使用枸橼酸钠和 $ZnCl_2$ 分别作为碳源和掺杂剂,通过水热法制备了锌掺杂碳点(Zn－CDs)。所得的 Zn－CDs 具有优异的荧光性质,并可以作为荧光探针用于分析检测[157]。此外,在 2012 年,Bourlinos 等人以三羟甲基氨基甲烷、甜菜碱盐酸盐和钆喷酸为前驱体制备了钆掺杂 CDs(Gd－CDs),获得的 Gd－CDs 显示出均匀的粒径并可以用于磁共振成像[158]。

1.5.1.6 多原子共掺杂

虽然,单原子掺杂已显示出调控 CDs 固有性质的巨大潜力,但同时也存在着一些局限性。对此,有许多报道表明可以通过同时掺杂两种或多种不同原子来突破单原子掺杂 CDs 的瓶颈,这称为多原子共掺杂。

氮和硫共掺杂 CDs(N, S－CDs)已被许多科研团队研究并报道,这主要是因为氮的原子半径接近碳原子,同时硫的电负性类似于碳原子,使得这种掺杂易于形成。首次报道成功制备的 N, S－CDs 是通过水热处理柠檬酸和 Cys 得到的,该 N, S－CDs 具有优异的荧光性能和低毒性,在生物成像中得到了应用[159]。同时,氮和磷共掺杂对 CDs 的极性、量子产率以及电化学性质具有极大影响。在 2016 年,Shi 等人通过水热法制备的氮和磷共掺杂 CDs(N, P－CDs)[160]。之后,他们将 N, P－CDs 作为一种新型荧光探针用于定量分析人血清和活细胞中的 Fe^{3+}。此外,研究发现将元素周期表中与碳相邻的硼和氮两种元素进行掺杂可以调整 CDs 的导带/价带位置,从而改变 CDs 的电导率和发光性能。2016 年,Tan 等人以 1,2－十六烷二醇为碳源,三聚氰胺和硼酸钠分别作为氮源和硼源,通过热解法制备了氮和硼共掺杂 CDs(N, B－CDs)。与 N－CDs 相比,N, B－CDs 展现了极高的量子产率。同时,N, B－CDs 的大小和表面态都会影响荧光性质。最后,将 N, B－CDs 用于传感和光学设备中[161]。除上述提及的几种类型的多原子共掺杂外,还存在钴铁共掺杂、氧氮共掺杂、氟氮共掺杂、氮镁共掺杂、氮磷硫共掺杂等[162-167]。

1.5.2 溶剂效应

反应溶剂对 CDs 的表面态/结构有很大影响,并且反应溶剂差异将引起 CDs 的光学性质发生重大变化。例如,Liu 课题组使用柠檬酸和 4 种含氮有机物为前驱体,甲苯和水作为反应溶剂,通过反应溶剂和前驱体的不同组合制备了 8 种 CDs,并系统地研究了反应溶剂对所制备 CDs 发光的影响(图 1-9)[168]。研究发现,以甲苯作为反应溶剂制备的所有 CDs 产物都产生 2 种不同颜色的荧光发射峰(蓝色和黄色),并且通过改变前驱体可以实现 2 种颜色发射峰相对强度的精确调节。但是,以水作为反应溶剂制备的 CDs 仅显示蓝色发射。这说明反应溶剂的选择对 CDs 固有的荧光发射调控具有极其重要的作用和影响。Zhang 等人通过调控反应溶剂的成分,制备出了尺寸、组成和官能团可控的多色荧光 CDs,并成功用于体外和体内成像[169]。此外,Xiong 等人以邻苯二胺和谷氨酸为碳源,通过改变反应溶剂成功制备了 8 种不同荧光发射的 CDs[170]。研究发现,通过改变反应溶剂可以改变 CDs 的尺寸和石墨氮的含量,进而影响 CDs 的发光性能。随后,他们将 CDs 分散到聚乙烯醇基质中制备了具有可调颜色的发光膜。同时基于 CDs 的高量子产率、良好的光稳定性和低毒性还将其用于体内外生物成像。

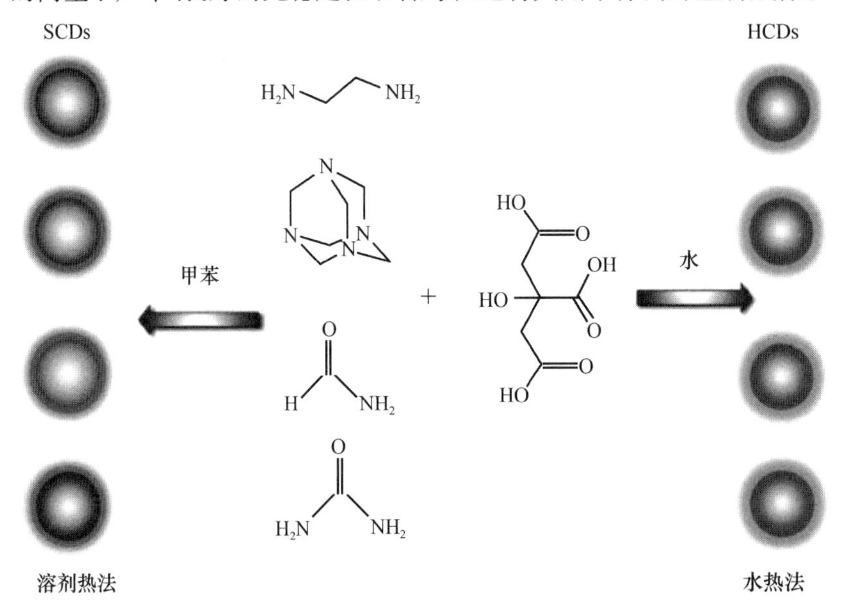

图 1-9　不同反应溶剂中 CDs 的制备过程示意图[168]

1.5.3 表面修饰

CDs 的荧光特性很大程度上会受到表面改性和钝化的影响。例如:Ding 等人通过水热法制备出了在单色激发波长下从蓝色到红色发光的 CDs[13]。研究发现,CDs 样品具有相似的粒径分布和石墨结构分布,但它们之间的表面态氧化程度逐渐变大。因此,可以观察到它们的发射峰从紫色区域移动到红色区域。此外,在某些情况下,表面未钝化的 CDs 可能没有荧光产生,而钝化后的 CDs 便可以检测到荧光发射[11,171]。因此,只有进一步将 CDs 表面改性或钝化处理才能进行应用。目前,各种用于改性或钝化 CDs 的修饰剂层出不穷,其中小分子和聚合物已被广泛使用。在 Tetsuka 课题组的研究中,为了保护石墨烯结构免遭破坏,他们在 70 ~ 150 ℃的氨溶液中对氧化石墨烯进行了水热处理,通过氨介导的氧化石墨烯切割工艺来设计光学可调的氨基官能化 CDs[172]。研究发现,较低的水热温度和较高的氨浓度可以引起较高的氮碳比。同时,通过控制伯胺官能化程度,可以将 CDs 的荧光发射从紫色区域变为黄色区域。此外,他们进一步评估了其他不同的含氮官能团对 CDs 的能级和带隙的影响(图 1 - 10)[173]。研究发现,氮成分与石墨烯核的强烈轨道共振将会引发 CDs 的 HOMO 和 LUMO 水平被不同的氮成分连续调节。具体而言,用邻苯二胺、二氨基萘、偶氮或对甲基红功能化的 CDs 具有较低的能级,而经胺或二甲胺修饰的 CDs 具有简并的 HOMO 轨道和较高的能级。Qiu 等人通过酰胺键桥连反应将各种有机氨分子连接在 CDs[174]。与未修饰 CDs 相比,修改后的 CDs 均显示出更高的量子产率。更重要的是,这些氨基分子的烃链长度可以很容易地调节 CDs 的表面极性。Gupta 等人使用类似的接枝方法用苯胺修饰了 CDs,所得 CDs 不仅荧光颜色从蓝色改变为白色,而且其极性也从亲水性改变为疏水性[175]。

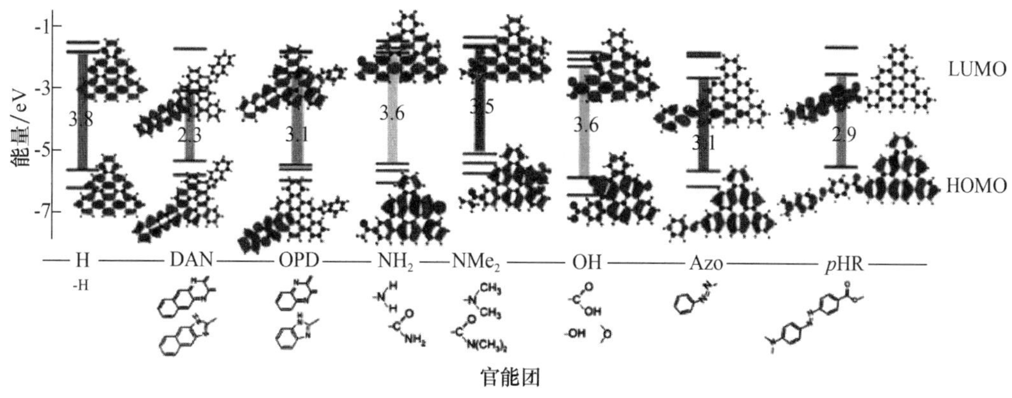

(a) 不同氮官能化CDs的能级和荧光

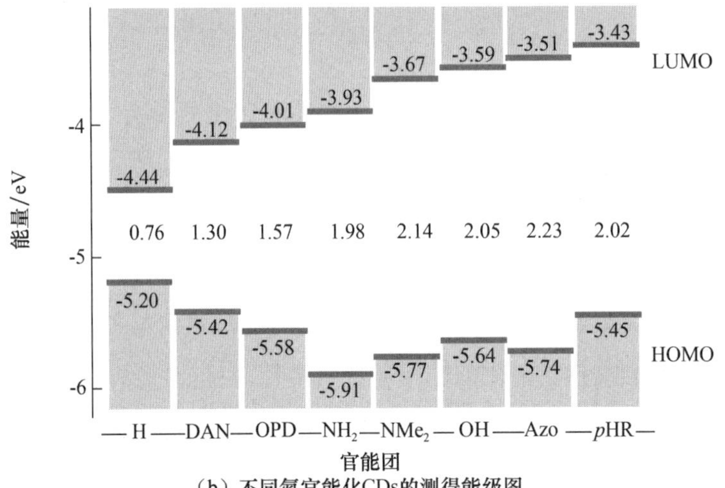

（b）不同氮官能化CDs的测得能级图

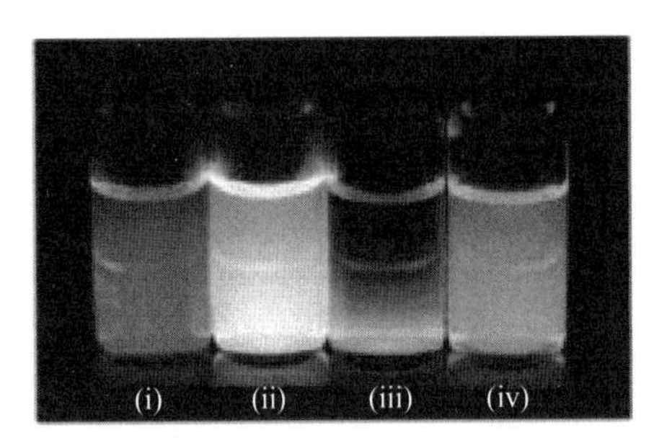

（c）使用紫外线灯（365 nm）照射的不同氮官能化CDs水溶液图像

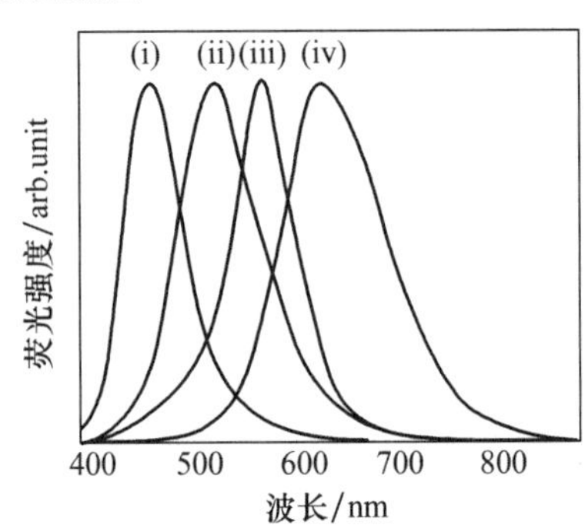

（d）不同氮官能化CDs水溶液的相应归一化荧光光谱（在380 nm处激发）

图 1-10

除了上述的表面态调控方法外，CDs 的吸收和发射性能也可以通过表面电荷进行调控[77]。通过与碱反应，制备的 CDs 被金属阳离子官能化，并表现出橙色发射。相反，在稀盐酸溶液中除去表面官能化的金属阳离子后，非金属阳离子官能化的 CDs 在吸收光谱和发射光谱之间显示出重叠的增加。此现象可能是因为表面金属阳离子官能化可提高 CDs 的费米能级，从而引起 CDs 自吸收降低和橙色发射增强。

1.6 碳点的应用

1.6.1 在光学分析领域的应用

正因为 CDs 具有低毒性、良好的光稳定性、优异的水分散性、强荧光性和磷光性等特点,所以它们在传感器、信息加密等光学领域得到了广泛的应用。

1.6.1.1 传感应用

CDs 由于其固有的荧光特性、高灵敏度、快速响应、成本低廉以及简单的制备方法,已被广泛用作检测环境或生物体系中各种分析物的荧光探针。小尺寸、大比表面积和丰富的表面官能团使 CDs 对周围环境(例如:温度、离子强度和溶剂)具有很高的响应性和灵敏性从而引起其特性(尤其是光学特性)发生变化,例如:增强/激活(打开)和淬灭(关闭)荧光。从理论上讲,基于 CDs 的荧光探针检测机理主要包括:光致电子转移、荧光共振能量转移以及内滤效应[4,176]。

通常,CDs 可以作为荧光探针用于检测阳离子和阴离子。检测物可以通过配位/静电相互作用与 CDs 表面基团结合[177-179]。Yang 等人以葡萄糖酸锌为碳源通过一步法制备了 Zn^{2+} 钝化的 CDs。通过与葡萄糖为碳源制备的 CDs 相比,他们提出 Zn^{2+} 可以防止碳化过程中的聚集,还可以钝化 CDs 的表面以进一步稳定 CDs 并提高其荧光强度。受到 Zn^{2+} 与乙二胺四乙酸的络合反应的启发,他们设计了一种可逆的荧光“开关”纳米传感器用于选择性检测乙二胺四乙酸和 Zn^{2+} [图 1-11(a)][180]。Chen 等人通过固定在医用棉布上的带有 pH 响应性橙色荧光 CDs,实现了荧光和可见比色双模式检测伤口 pH 值[图 1-11(b)][181]。此外,该检测方法还可以在血液污染和长期观察的情况下,从理论上和视觉上估算伤口的 pH 值。除了在离子和 pH 传感方面的应用外,CDs 还显示出对多种生物分子的选择性,包括:氨基酸、谷胱甘肽、维生素、甲醛、葡萄糖、DNA 和蛋白质等一些与健康相关的分子[182-190]。因此,CDs 可以作为探针为疾病的早期预防和诊断提供有价值的参考。

由于低毒性、出色的生物相容性和光稳定性,CDs 还作为探针用于癌细胞的有效靶向和成像中。同时,还凭借修饰配体方法用于鉴定和检测细菌。例如,Gao 等人报道了一种具有线粒体靶向的绿光荧光 CDs,由于此 CDs 在线粒体膜电位和物质吸收效率上表现的差异性,可以有效地将癌细胞与正常细胞区分开[图 1-11(c)][191]。Mao 等人制备了一种具有优异光学性能的正电荷 N,S-CDs,并通过对 CDs 进行修饰适配体的方法,使 CDs 展现出对靶细胞的高亲和力和特异性[192]。Yang 课题组设计了一种蓝色和红色荧光的嗜酸 CDs,通过环境和细菌的细胞壁区别,实现了对不同细菌的区分[193]。与牙龈卟啉单胞菌相比,通过 CDs 标记后变形链球菌显示出更强的绿色和红色荧光,同时轮廓更加清晰。而与之相比,金黄色葡萄球菌和大肠杆菌仅呈现单一颜色荧光[图 1-11(d)]。这项工作将 CDs 设计成了一种新颖的造影剂。与传统方法相比,CDs 不但可以对细菌进行成像,而且还可以清晰地识别并区分细菌。

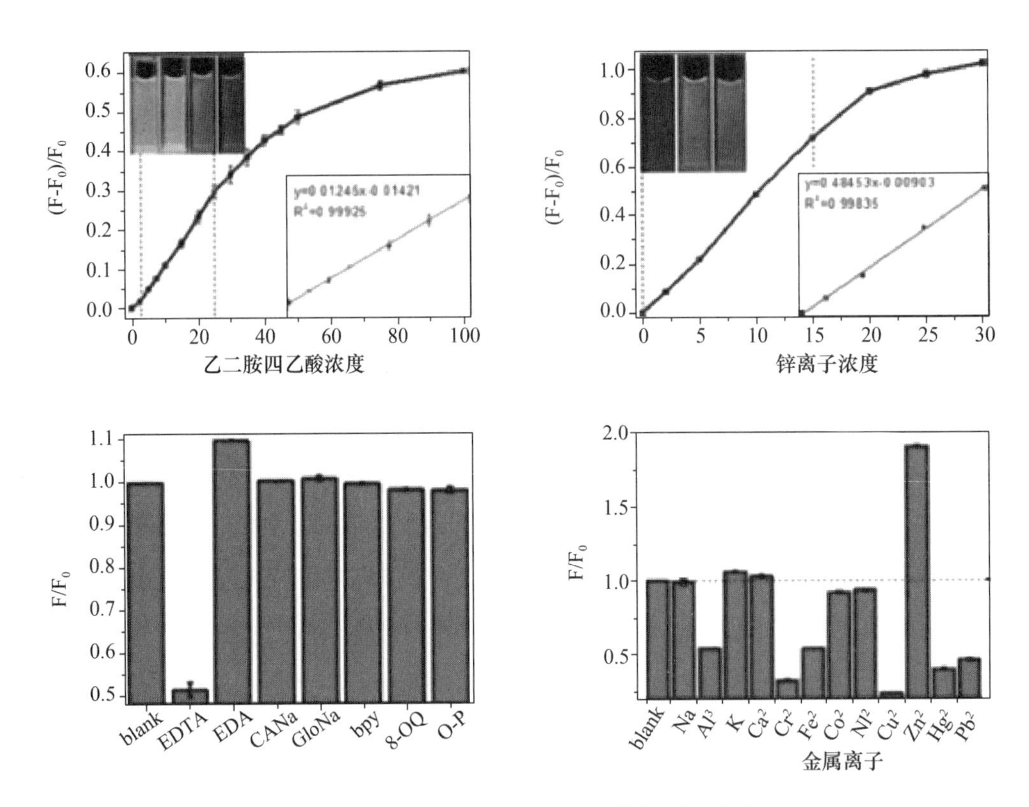

(a) CDs在EDTA和Zn^{2+}传感中的应用[180]

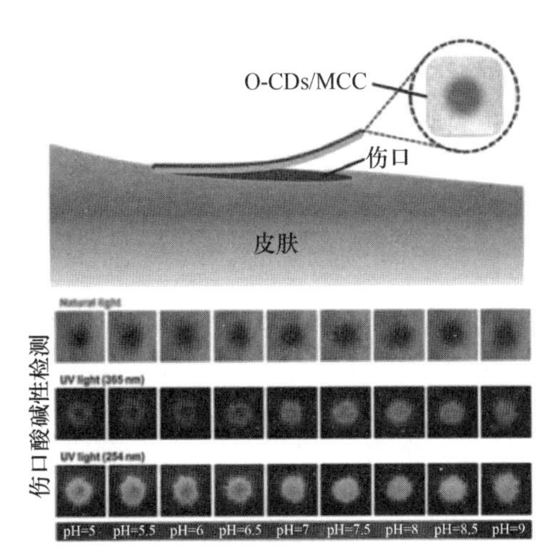

(b) 与医用棉布结合的CDs在不同pH值下的实际应用和图像[181]

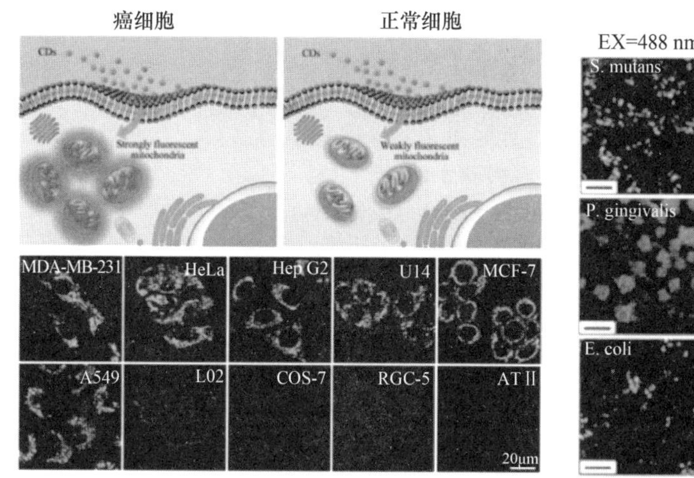

（c）通过CDs将癌细胞与正常细胞区分开[191]

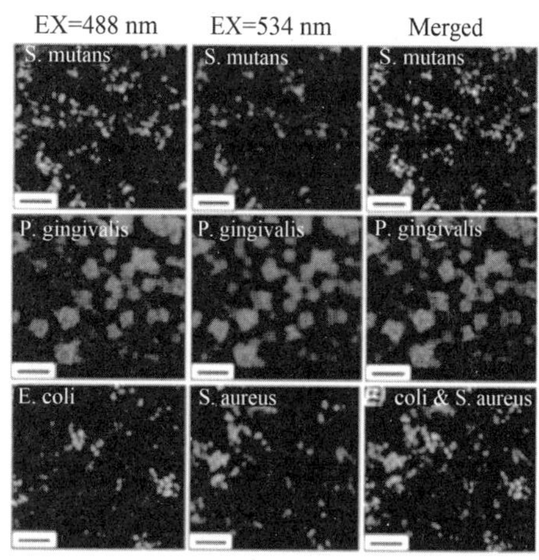

（d）嗜酸性CDs通过荧光图像区分4种细菌[193]

图1-11 CDs的传感应用

1.6.1.2 信息加密

信息加密有助于保护有价值的东西不被复制。与传统的防伪技术或具有室温磷光的有机金属配合物（含金属）和纯有机化合物（不含金属）的材料相比，CDs具有环保、易于处理、易于设计以及成本低廉等特点。这也使CDs在数据防伪和加密领域具有很大的应用前景。

最初，将CDs用于信息加密领域是将CDs嵌入到各种基质中，以获得基于CDs的RTP材料。但是，这些主体仅允许特定的客体CDs生成RTP，缺乏通用性。Li课题组报道了一种通过硼酸基质激活多色RTP-CDs进行防伪技术和信息加密的通用方法[图1-12(a)][22]。然而，这种复合物中可逆相互作用的不稳定性限制了它们的实际应用。为了解决上述问题，Yang课题组设计了由聚丙烯酸和乙二胺制备的新型无金属RTP-CDs，而无须另外的基质[21]。此外，图像证明了该CDs在安全保护方面具有很广阔的应用前景[图1-12(b)]。最近的研究表明，具有RTP和热活化延迟荧光的氧化控制CDs系统可以进一步保证防伪和多层信息安全[25]。这些对CDs结构设计以及光物理性质的研究为开发高性能余辉发光材料提供了新的方向。

（a）多色CDs用于数字信息加密[22]

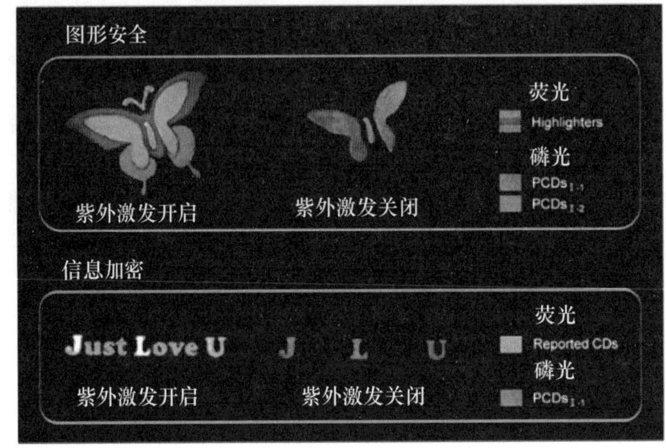

（b）CDs用于图形安全和数字信息加密[21]

图 1－12　CDs 用于信息加密

1.6.2　在能源领域的应用

加速的工业发展和经济增长导致化石燃料的快速消费、能源短缺、环境污染和气候变化等问题。因此，探索和开发可再生、环境友好、稳定和高效的能源转换和存储技术对于全社会发展至关重要且迫在眉睫[194－196]。作为一种新兴的碳材料，CDs 由于其可调节的光学性质、低成本、低毒性、大比表面积、出色的电子受体/供体特性以及电子传导性等有望在光/电催化剂、发光二极管、超级电容器、可充电电池、金属空气电池和燃料电池等能源领域得到广泛应用。

1.6.2.1　催化应用

基于 CDs 的不同结构和性质，目前已经提出了将 CDs 作为光催化剂、电催化剂和光电催化剂用于催化领域。

太阳能是一种可再生的清洁能源。为了充分利用太阳能，光催化进入人们的视野，并成了一种引人注目的应用。在太阳能转化过程中，CDs 可以作为光催化剂用于污染物降解、水分解、CO_2 还原和化学反应[197－205]。这得益于 CDs 从紫外到近红外光吸收范围、出色的光稳定性以及电荷分离转移能力。CDs 中位于芳族区域边缘位置的氮原子能够有效地进行界面电子转移,并进一步通过光催化作用增加水中氢的产生[206]。通常,与其他纳米材料(Fe_2O_3, g－C_3N_4)复合的 CDs 在异质结中充当吸光剂或电子受体以提高光催化性能[200－202,204,207]。例如,与 CDs 耦合的赤铁矿展现了高效的光催化 H_2 和 O_2 的释放速率,这是因为 CDs 产生的光致电势,增强了 Fe_2O_3 的光吸收并改善电荷分离效率[208]。Yang 课题组发现,将 CDs 整合到 g－C_3N_4 中有助于在层间/层内运输载流子并能够抑制重组,从而显著提高罗丹明 B 的光催化降解效率,提升 CO_2 还原和水分解的产生 H_2 的光催化速率,这项研究成功将环境修复与能量转换结合起来[198]。

1.6.2.2 发光二极管应用

发光二极管(LED)是一种固态照明设备,它能将电能转换为光能。多年来,它一直是学术研究的热门课题。人们对其研究的最终目的是将其应用于日常生活中的液晶显示器、全彩显示器和照明设备中。作为一种新兴的荧光材料,CDs 因为具有丰富的色光、可调节的荧光颜色、低成本且对环境友好,所以可以取代 LED 中昂贵的稀土基磷光体和有毒金属基半导体量子点。通常,CDs 可以充当 CDs 基 LED(CLED)中的磷光体或电致发光器件中的有源层。

在基于磷光体的 CLED 中,大多数 CDs 均匀地分散在基质材料中以防止由于过度的共振能量转移或直接的 π－π 相互作用而导致的固态/粉末状态下 CDs 的聚集诱导荧光淬灭[209－213]。目前,通过这些方法多色/白色 CLED 已经被成功制备。然而,这些常用方法需要烦琐的处理过程,并且厚重的固体基质层的引入会抑制芯片的发光并增加成本。近年来,以聚合物、无机盐以及硅烷偶联剂为基础的一些抗自熄性固态荧光 CDs 被开发并直接作为 CLED 中的磷光体,大大简化了制备过程并降低了成本[47,214,215]。然而,截至目前,获得具有高发光效率和显色指数的多色/白色 CLED 仍然是个不小的挑战。

在电致发光 CLED 中,CDs 通常用作夹在多层器件结构中的有源发射层。Ma 等人首先提出了将柠檬酸衍生的 CDs 用于白色 CLED 器件的发射层,其最大亮度和外部量子效率分别达到 35 cd m^{-2} 和 0.083%[216]。受这项工作的启发,最近在电致发光 CLED 应用方面取得了一些重大进展。例如,Yuan 课题组采用高性能的多色三角 CDs 作为发光层来制备多色 CLED,这个 CLED 具有极高的色纯度,最大亮度为 1882－4762 cd m^{-2},电流效率为 1.22－5.11 cd A^{-1}[217]。随后,他们通过优化 CDs 的制备过程,报道了一种深蓝色 CLED,其性能远胜于基于 Cd^{2+}/Pb^{2+} 的材料的深蓝色 LED[218]。尽管如此,与成熟的基于 Cd^{2+}－量子点的 LED(QLED)或出色的钙钛矿 QLED 相比,CLED 的研究仍处于早期阶段。CLED 在亮度、外部量子效率等参数方面还有一定的提升空间。

1.6.2.3 充电电池

充电电池被认为是架桥可再生能源生产和消耗的最有效的能量存储技术之一。在 Li、Na 和 K 离子电池中,CDs 可以通过表面工程为电极和电解质之间创建界面,为离子插入和提取提供更多的活性位点,有助于提高稳定性、增强电子/离子的传输和扩散并改善电化学性能[219-224]。CDs 已被作为“设计添加剂”用于富含石墨烯的花瓣状金红石型 TiO_2 中,以缩短 Na^+ 的扩散途径并提高集合体的电导率,从而使钠储存具有高容量和长期循环稳定性[219]。此外,CDs 的含氧官能团有助于吸引金属阳离子,产生均匀的固体电解质界面,并保持电极结构的完整性,从而改善电化学性能[220]。为了降低多硫化物穿梭的影响,增加面硫负荷,并增大锂硫电池的工作电流密度。Xiong 等设计了聚乙烯亚胺官能化的 CDs 修饰的阴极[225]。实验结果表明 CDs 的小尺寸、出色的分散性和更多的氨基基团能够提供丰富的吸收位点,可以有效地抑制多硫化物的溶解并增加固体-电解质界面周围的 Li^+ 电导率。

1.6.3 在生物领域的应用

在 CDs 的所有应用中,生物医学是最有前途和最常被报道的应用领域之一。通过对一系列细胞系的体外细胞毒性研究发现,CDs 即使在高浓度水平下,也表现出了低毒或无毒的特性,并且展现了出色的生物相容性[2,12,226-230]。同时,体内实验表明 CDs 可以通过肾脏或肝胆系统迅速排泄。此外,根据血液生化和血液学分析,在小鼠的脑、心脏、肺、肝、脾、肾、睾丸和膀胱中未观察到明显的炎症症状[2,31]。因此,CDs 在生物医学领域应用是安全、可靠的。低成本、小尺寸、可控制的表面功能、极高的光稳定性、独特的荧光特性以及高亮度使 CDs 有望成为疾病诊断治疗中替代传统荧光材料的新型纳米材料。

1.6.3.1 生物成像

生物成像是一种通过探针和检测器进行实时、无创直接可视化生物动态的技术。荧光生物成像作为一种成像方式,由于其便利性、低成本、高灵敏度、无创性和长期观察性,已成为一种强大的临床诊断方法。但是,传统的荧光团(例如,量子点和有机染料)存在毒性或荧光性能差等问题。鉴于 CDs 的高光稳定性、出色的生物相容性、简单的制备路线、灵活的可设计性、多色发射、深红色/近红外发射以及两光子/多光子荧光等特性,CDs 自然而然地成为下一代体外和体内生物成像荧光探针的主要研究对象。

目前,许多种类的 CDs 已被设计用于细胞、微生物和植物组织的成像[193,231-237]。通常,根据不同的 CDs 结构和不同的细胞类型,CDs 可以通过能量/温度依赖的细胞结构的内吞作用进入细胞,并分布到不同种类的细胞器中[238-244]。细胞器成像对于了解和研究与细胞器相关的疾病很有帮助。Wu 等人发现用间苯二胺和 Cys 制备的 CDs 可以用于活细胞核的成像[243]。此外,与原卟啉 IX 结合后,CDs 获得了细胞核靶向的光动力疗法能力,从而在激光照射后无毒性影响的情况下有效消融肿瘤。植物乳杆菌衍生的 CDs 作为染色剂用作微生物成像,这可以提供与细菌的形态和生理状态相关的有价值的信息[233]。

除了体外成像外,组织自发荧光和光散射的现象为成像提供了出色的对比度和空间分辨率,因此具有红色/近红外发射或两光子/多光子荧光的 CDs 具有成为体内荧光跟踪剂的潜力。Qu 等人通过表面工程设计了由两个和三个光子诱导的深红色发射 CDs,并将其用于活体小鼠胃的体内深红色荧光成像[15]。最近,Yang 课题组报道了量子产率高达 59% 的深红色发射 CDs 作为一种单光子和两光子生物成像的有效探测器[2]。他们通过实时体内成像系统地研究了 CDs 在裸鼠体内的生物分布情况。结果表明,CDs 可以在几分钟内通过血液循环迅速进入小鼠全身。首先,CDs 大量积累在肝、肺和肾中,然后在 24 小时内通过肾脏和肝胆系统逐渐清除。此外,他们发现用 o-PD 制备的红色荧光 CDs 在无须靶向剂(例如,转铁蛋白)情况下便可轻松穿过健康小鼠的血脑屏障,这为实时跟踪预防和治疗脑部疾病提供了新的材料[31]。

1.6.3.2　药物和基因运输

除了用于细胞成像外,CDs 可以将成像与药物和基因运输相结合,形成成像参与的纳米杂交体,从而提高疾病治疗的递送效率。

药物输送是一种安全、有效的治疗方法,是指将药物运送到体内的特定位置并持续释放的过程。因此,在药物递送系统中受控的药物释放和强大的选择性对于增加局部治疗效果以及健康组织的副作用最小化至关重要。CDs 通过其荧光特性在可视化病理部位的药物蓄积和活性方面具有优势,这对于评估药物的治疗效果至关重要[245-249]。通过追踪叶酸修饰的 CDs 与氯喹结合后的绿色荧光,Fan 课题组实现了肿瘤治疗的实时成像和监控[245]。同样地,将 CDs 与抗癌药物多西紫杉醇包裹在纳米海绵中能够有效地进行药物输送、成像和对深层肿瘤的光解[246]。为了改善肿瘤特异性成像和药物传递性能,Zhou 等人设计了一种深红色发射 CDs,因为它们与转运蛋白的多价相互作用,所以可以靶向包括神经胶质瘤在内的肿瘤组织。最终,装载了盐酸拓扑替康的 CDs 可用于荧光成像和疾病的治疗[247]。

与其他治疗技术不同,基因治疗被认为是针对各种疾病的持久且可能有效的临床方法[250,251]。基因治疗中有效的载体可以将遗传物质传递到细胞中,并具有很高的基因转染效率。具有天然侵入能力和传递其遗传物质的病毒载体已成为有效的基因载体。但是,由于潜在的严重安全隐患使其在临床使用上受到了限制[251]。CDs 具有低毒性、丰富的官能团和出色的生物相容性。重要的是,CDs 的小尺寸有助于细胞对载体的充分摄取,从而提高了基因转染效率。此外,它们独特的荧光可用于追踪基因的内在化[250,251]。因此,CDs 在基因治疗中作为非病毒载体将具有极大的吸引力。

1.7　本学术专著选题意义及研究内容

近年来,CDs 作为荧光纳米材料家族的一个新成员,由于其杰出的光学性、水溶性、生物相容性等特点受到了人们的关注。正是由于这些优异的性质,也使得 CDs 在传感、生物成像、药物运输以及光催化等领域得到了广泛的应用和发展。

迄今为止,基于苯二胺的 CDs 的研究在制备策略、性能、机理和应用等方面都取得了很大进展。这些研究结果证实苯二胺衍生 CDs 可以为化学、生物学等交叉学科的基础研究提供新工具、新方法。然而,由于碳源、制备方法和提纯方法的多样性,关于不同表面态在 CDs 方面的研究仍然存在诸多疑问。例如:

(1)不同反应溶剂如何影响 CDs 的表面态;

(2)如何绿色制备不同表面态的 CDs;

(3)表面态差异如何影响 CDs 发光性。

因此,为了解决上述问题,本文从分子融合角度出发,通过反应溶剂参与的分子融合法、室温氧化融合法以及氟掺杂的表面态调控法制备了一系列不同表面态 o－PD 衍生 CDs,并系统地研究了这些 CDs 的形成过程和发光机理,探索所制备的 CDs 在化学传感以及生物成像领域的应用。具体研究内容包括以下三个方面:

(1)采用反应溶剂参与的分子融合法设计制备了一系列不同表面态 o－PD 衍生 CDs。其中以 o－PD 为碳源,甲酰胺为反应溶剂制备的 FCDs 为主要研究对象,提出了基于席夫碱反应的 FCDs 形成机理。同时,研究发现反应溶剂与前驱体分子之间的融合能够影响所制备 CDs 的表面态和结构,进而改善 CDs 的选择性。以 FCDs 为例,它可以作为荧光探针以荧光“开关”的方式用于 Ag^+ 和 Cys 的分析检测。此外,由于 FCDs 具有长波长发射、良好的生物相容性和较低的细胞毒性,FCDs 还可以作为探针用于 Ag^+ 和

Cys 的细胞成像，同时能够防止活体组织自体荧光的干扰。

（2）采用室温氧化融合法，在室温下以 o－PD 和 HQ 为原料，通过前驱体间的氧化、聚合及席夫碱反应一步法绿色制备了两种不同表面态的 o－PD 衍生 CDs（YCDs 和 GCDs）。基于两种 CDs 不同的极性，通过硅胶柱层析将 CDs 进行分离，并对它们的荧光发射机理进行研究。最后，根据荧光性质的不同，将 YCDs 作为荧光探针用于有毒污染物 p－NP 的检测，将 GCDs 用于分析 D_2O 中 H_2O 的含量。

（3）采用氟掺杂的表面态调控法，以 o－PD、p－BQ 以及 o－PD 衍生物 4－氟－1，2－苯二胺为碳源，在室温下制备了不同表面态的 o－PD 衍生 CDs（UCDs、FCDs1 和 FCDs2）。研究发现，通过氟含量控制可以有效调控 CDs 的荧光发射情况。基于不同的荧光发射特性，将制备的 FCDs1 作为荧光可视化染料用于潜指纹显现，FCDs2 作为荧光探针用于 CBL 的定量检测。

（4）采用简单的溶剂热法，以 o－PD 和乙醇为原料，制备出最大发射波长为 555 nm 的黄光氮掺杂 CDs（YNCDs）。基于内滤效应，姜黄素可以有效抑制 YNCDs 的荧光发射。因此，以 YNCDs 为纳米探针，建立了检测姜黄素的传感器平台，其检出限（3S/N）为 0.01 $\mu mol \cdot L^{-1}$。此外，该传感器平台实现了对咖喱粉、人尿、人血清等真实样品中姜黄素的高选择性、高灵敏度检测，准确性和回收率也令人满意。进一步地，YNCDs 优异的光学性能还可作为隐形油墨应用于信息存储和防伪等领域。

（5）采用简便、快捷的水热法，以 o－PD 为原料，制备了强黄光 CDs（SYCDs）。结晶紫作为一种危险染料，对环境和人类健康构成严重威胁。因此，本章开发了一种简便的方法来对结晶紫进行检测。研究发现，结晶紫可以有效抑制 SYCDs 的荧光发射。因此，以 SYCDs 为纳米探针，建立了检测结晶紫的传感器平台，当结晶紫浓度在 0.02 ～ 15 $\mu mol \cdot L^{-1}$范围内，具有良好线性规律，其检出限（3S/N）为 0.006 $\mu mol \cdot L^{-1}$。通过详细的研究，提出了内滤效应是 SYCDs 对结晶紫的传感机理。此外，该传感器平台实现了鱼组织真实样品中结晶紫的高选择性、高灵敏度检测，准确性和回收率也令人满意。

参考文献

[1] Yan Y, Gong J, Chen J, et al. Recent advances on graphene quantum dots: From chemistry and physics to applications [J]. Advanced Materials, 2019, 31 (21): 1808283.

[2] Liu J, Geng Y, Li D, et al. Deep red emissive carbonized polymer dots with unprecedented narrow full width at half maximum [J]. Advanced Materials, 2020, 32

(17): 1906641.

[3]Hu C, Li M, Qiu J, et al. Design and fabrication of carbon dots for energy conversion and storage [J]. Chemical Society Reviews, 2019, 48 (8): 2315 -2337.

[4]Liu M L, Chen B B, Li C M, et al. Carbon dots: Synthesis, formation mechanism, fluorescence origin and sensing applications [J]. Green Chemistry, 2019, 21 (3): 449 -471.

[5]Zhu S, Song H, Zhao X, et al. The photoluminescence mechanism in carbon dots (graphene quantum dots, carbon nanodots, and polymer dots): Current state and future perspective [J]. Nano Research, 2015, 8: 355 -381.

[6]Li T, Shi W, E S, et al. Green preparation of carbon dots with different surface states simultaneously at room temperature and their sensing applications [J]. Journal of Colloid and Interface Science, 2021, 591: 334 -342.

[7]Wang J, Li Q, Zheng J, et al. N, B -codoping induces high -efficiency solid -state fluorescence and dual emission of yellow/orange carbon dots [J]. ACS Sustainable Chemistry & Engineering, 2021, 9 (5): 2224 -2236.

[8]Baker S N, Baker G A. Luminescent carbon nanodots: Emergent nanolights [J]. Angewandte Chemie International Edition, 2010, 49 (38): 6726 -6744.

[9]Zhang W, Zhong H, Zhao P, et al. Carbon quantum dot fluorescent probes for food safety detection: Progress, opportunities and challenges [J]. Food Control, 2022, 133: 108591.

[10]Li T, Dong Y, Bateer B, et al. The preparation, optical properties and applications of carbon dots derived from phenylenediamine [J]. Microchemical Journal, 2023, 185: 108299.

[11]Xu X, Ray R, Gu Y, et al. Electrophoretic analysis and purification of fluorescent single -walled carbon nanotube fragments [J]. Journal of the American Chemical Society, 2004, 126: 12736 -12737.

[12]Sun Y -P, Zhou B, Lin Y, et al. Quantum -sized carbon dots for bright and colorful photoluminescence [J]. Journal of the American Chemical Society, 2006, 128: 7756 -7757.

[13]Zhu S, Meng Q, Wang L, et al. Highly photoluminescent carbon dots for multicolor patterning, sensors, and bioimaging [J]. Angewandte Chemie International Edition,

2013, 52 (14): 3953 -3957.

[14]Ding H, Yu S - B, Wei J - S, et al. Full - color light - emitting carbon dots with a surface - state - controlled luminescence mechanism [J]. ACS Nano, 2016, 10 (1): 484 -491.

[15]Miao X, Qu D, Yang D, et al. Synthesis of carbon dots with multiple color emission by controlled graphitization and surface functionalization [J]. Advanced Materials, 2018, 30 (1): 1704740.

[16]Li D, Jing P, Sun L, et al. Near - infrared excitation/emission and multiphoton - induced fluorescence of carbon dots [J]. Advanced Materials, 2018, 30 (13): 1705913.

[17]Lu S, Sui L, Liu J, et al. Near - infrared photoluminescent polymer - carbon nanodots with two - photon fluorescence [J]. Advanced Materials, 2017, 29 (15): 1603443.

[18]Shao J, Zhu S, Liu H, et al. Full - color emission polymer carbon dots with quench - resistant solid - state fluorescence [J]. Advanced Science, 2017, 4 (12): 1700395.

[19]Bao X, Yuan Y, Chen J, et al. In vivo theranostics with near - infrared - emitting carbon dots - highly efficient photothermal therapy based on passive targeting after intravenous administration [J]. Light: Science & Applications, 2018, 7: 91.

[20]Ye X, Xiang Y, Wang Q, et al. A red emissive two - photon fluorescence probe based on carbon dots for intracellular pH detection [J]. Small, 2019, 15 (48): 1901673.

[21]Li F, Li Y, Yang X, et al. Highly fluorescent chiral N - S - doped carbon dots from cysteine: Affecting cellular energy metabolism [J]. Angewandte Chemie International Edition, 2018, 57 (9): 2377 -2382.

[22]Tao S, Lu S, Geng Y, et al. Design of metal - free polymer carbon dots: A new class of room - temperature phosphorescent materials [J]. Angewandte Chemie International Edition, 2018, 57 (9): 2393 -2398.

[23]Li W, Zhou W, Zhou Z, et al. A universal strategy for activating the multicolor room - temperature afterglow of carbon dots in a boric acid matrix [J]. Angewandte Chemie International Edition, 2019, 58 (22): 7278 -7283.

[24]Xia C, Tao S, Zhu S, et al. Hydrothermal addition polymerization for ultrahigh - yield carbonized polymer dots with room temperature phosphorescence via nanocomposite [J]. Chemistry - A Europcan Journal, 2018, 24 (44): 11303 - 11308.

[25] Jiang K, Hu S, Wang Y, et al. Photo - stimulated polychromatic room temperature phosphorescence of carbon dots [J]. Small, 2020, 16 (31): 2001909.

[26] Park M, Kim H S, Yoon H, et al. Controllable singlet - triplet energy splitting of graphene quantum dots through oxidation: From phosphorescence to TADF [J]. Advanced Materials, 2020, 32 (31): 2000936.

[27] Liu J, Li R, Yang B. Carbon dots: A new type of carbon - based nanomaterial with wide applications [J]. ACS Central Science, 2020, 6 (12): 2179 - 2195.

[28] Pan D, Zhang J, Li Z, et al. Hydrothermal route for cutting graphene sheets into blue-luminescent graphene quantum dots [J]. Advanced Material, 2010, 22 (6): 734 - 738.

[29] Zhu S, Zhang J, Tang S, et al. Surface chemistry routes to modulate the photoluminescence of graphene quantum dots: From fluorescence mechanism to up - conversion bioimaging applications [J]. Advanced Functional Materials, 2012, 22 (22): 4732 - 4740.

[30] Peng J, Gao W, Gupta B K, et al. Graphene quantum dots derived from carbon fibers [J]. Nano Letters, 2012, 12 (2): 844 - 849.

[31] Li L, Wu G, Yang G, et al. Focusing on luminescent graphene quantum dots: Current status and future perspectives [J]. Nanoscale, 2013, 5 (10): 4015 - 4039.

[32] Liu J, Li D, Zhang K, et al. One - step hydrothermal synthesis of nitrogen - doped conjugated carbonized polymer dots with 31% efficient red emission for in vivo imaging [J]. Small, 2018, 14 (15): 1703919.

[33] Zhu S, Song Y, Shao J, et al. Non - conjugated polymer dots with crosslink - enhanced emission in the absence of fluorophore units [J]. Angewandte Chemie International Edition, 2015, 54: 14626 - 14637.

[34] Bourlinos A B, Stassinopoulos A, Anglos D, et al. Photoluminescent carbogenic dots [J]. Chemistry of Materials, 2008, 20 (14): 4539 - 4541.

[35] Pan D, Zhang J, Li Z, et al. Blue fluorescent carbon thin films fabricated from dodecylamine - capped carbon nanoparticles [J]. Journal of Materials Chemistry, 2011, 21 (11): 3565 - 3567.

[36] Zheng M, Ruan S, Liu S, et al. Self - targeting fluorescent carbon dots for diagnosis of brain cancer cells [J]. ACS Nano, 2015, 9 (11): 11455 - 11461.

[37] Jia Q, Ge J, Liu W, et al. A magnetofluorescent carbon dot assembly as an acidic

$H2_0^{2-}$ driven oxygenerator to regulate tumor hypoxia for simultaneous bimodal imaging and enhanced photodynamic therapy [J]. Advanced Materials, 2018, 30 (13): 1706090.

[38] Pei X, Xiong D, Wang H, et al. Reversible phase transfer of carbon dots between an organic phase and aqueous solution triggered by CO_2 [J]. Angewandte Chemie International Edition, 2018, 57 (14): 3687 – 3691.

[39] Wu Z L, Liu Z X, Yuan Y H. Carbon dots: Materials, synthesis, properties and approaches to long – wavelength and multicolor emission [J]. Journal of Materials Chemistry B, 2017, 5 (21): 3794 – 3809.

[40] Sun Y – P, Zhou B, Lin Y, et al. Quantum – sized carbon dots for bright and colorful photoluminescence [J]. Journal of the American Chemical Society, 2006, 128 (24): 7756 – 7757.

[41] Liu S, Tian J, Wang L, et al. Hydrothermal treatment of grass: A low – cost, green route to nitrogen – doped, carbon – rich, photoluminescent polymer nanodots as an effective fluorescent sensing platform for label – free detection of Cu(II) ions [J]. Advanced Materials, 2012, 24 (15): 2037 – 2041.

[42] Li T, Shi W, Mao Q, et al. Regulating the photoluminescence of carbon dots via a green fluorine – doping derived surface – state – controlling strategy [J]. Journal of Materials Chemistry C, 2021, 9: 17357 – 17364.

[43] Yang Y, Zhang H, Chen J, et al. A phenylenediamine – based carbon dot – modified silica stationary phase for hydrophilic interaction chromatography [J]. Analyst, 2020, 145 (3): 1056 – 1061.

[44] Qu D, Sun Z. The formation mechanism and fluorophores of carbon dots synthesized via a bottom – up route [J]. Materials Chemistry Frontiers, 2020, 4 (2): 400 – 420.

[45] Kanwa N, M K, Chakraborty A. Discriminatory Interaction Behavior of Lipid Vesicles toward Diversely Emissive Carbon Dots Synthesized from Ortho, Meta, and Para Isomeric Carbon Precursors [J]. Langmuir, 2020, 36 (35): 10628 – 10637.

[46] Jiang K, Sun S, Zhang L, et al. Red, Green, and Blue Luminescence by Carbon Dots: Full – Color Emission Tuning and Multicolor Cellular Imaging [J]. Angewandte Chemie International Edition, 2015, 54 (18): 5360 – 5363.

[47] Li H, He X, Kang Z, et al. Water – soluble fluorescent carbon quantum dots and pho-

tocatalyst design [J]. Angewandte Chemie International Edition, 2010, 49 (26): 4430 - 4434.

[48] Liu E, Li D, Zhou X, et al. Highly emissive carbon dots in solid state and their applications in light - emitting devices and visible light communication [J]. ACS Sustainable Chemistry & Engineering, 2019, 7 (10): 9301 - 9308.

[49] Yang S - T, Wang X, Wang H, et al. Carbon dots as nontoxic and high - performance fluorescence imaging agents [J]. The Journal of Physical Chemistry C, 2009, 113: 18110 - 18114.

[50] Wang R, Lu K - Q, Tang Z - R, et al. Recent progress in carbon quantum dots: Synthesis, properties and applications in photocatalysis [J]. Journal of Materials Chemistry A, 2017, 5 (8): 3717 - 3734.

[51] Hu S, Liu J, Yang J, et al. Laser synthesis and size tailor of carbon quantum dots [J]. Journal of Nanoparticle Research, 2011, 13 (12): 7247 - 7252.

[52] Xu H, Yan L, Nguyen V, et al. One - step synthesis of nitrogen - doped carbon nanodots for ratiometric pH sensing by femtosecond laser ablation method [J]. Applied Surface Science, 2017, 414: 238 - 243.

[53] Lu J, Yang J - X, Wang J, et al. One - pot synthesis of fluorescent carbon nanoribbons, nanoparticles, and graphene by the exfoliation of graphite in ionic liquids [J]. ACS Nano, 2009, 3: 2367 - 2375.

[54] Zhou J, Booker C, Li R, et al. An electrochemical avenue to blue luminescent nanocrystals from multiwalled carbon nanotubes (MWCNTs) [J]. Journal of the American Chemical Society, 2007, 129: 744 - 745.

[55] Liu M, Xu Y, Niu F, et al. Carbon quantum dots directly generated from electrochemical oxidation of graphite electrodes in alkaline alcohols and the applications for specific ferric ion detection and cell imaging [J]. Analyst, 2016, 141 (9): 2657 - 2664.

[56] Qiao Z - A, Wang Y, Gao Y, et al. Commercially activated carbon as the source for producing multicolor photoluminescent carbon dots by chemical oxidation [J]. Chemical Communications, 2010, 46 (46): 8812 - 8814.

[57] Bao L, Liu C, Zhang Z L, et al. Photoluminescence - tunable carbon nanodots: Surface - state energy - gap tuning [J]. Advanced Materials, 2015, 27 (10): 1663 - 1667.

[58] Xu Z - Q, Yang L - Y, Fan X - Y, et al. Low temperature synthesis of highly stable

phosphate functionalized two color carbon nanodots and their application in cell imaging [J]. Carbon, 2014, 66: 351 - 360.

[59] Li H, Liu R, Liu Y, et al. Carbon quantum dots/Cu_2O composites with protruding nanostructures and their highly efficient (near) infrared photocatalytic behavior [J]. Journal of Materials Chemistry, 2012, 22 (34): 17470 - 17475.

[60] Li H, He X, Liu Y, et al. Synthesis of fluorescent carbon nanoparticles directly from active carbon via a one - step ultrasonic treatment [J]. Materials Research Bulletin, 2011, 46 (1): 147 - 151.

[61] Park S Y, Lee H U, Park E S, et al. Photoluminescent green carbon nanodots from food - waste - derived sources: Large - scale synthesis, properties, and biomedical applications [J]. ACS Applied Materials & Interfaces, 2014, 6 (5): 3365 - 3370.

[62] Zhuo S, Shao M, Lee S - T. Upconversion and downconversion fluorescent graphene quantum dots: ultrasonic preparation and photocatalysis [J]. ACS Nano, 2012, 6 (2): 1059 - 1064.

[63] Wang Q, Liu X, Zhang L, et al. Microwave - assisted synthesis of carbon nanodots through an eggshell membrane and their fluorescent application [J]. Analyst, 2012, 137 (22): 5392 - 5397.

[64] Li F, Li C, Liu J, et al. Aqueous phase synthesis of upconversion nanocrystals through layer - by - layer epitaxial growth for in vivo X - ray computed tomography [J]. Nanoscale, 2013, 5 (15): 6950 - 6959.

[65] Zhu H, Wang X, Li Y, et al. Microwave synthesis of fluorescent carbon nanoparticles with electrochemiluminescence properties [J]. Chemical Communications, 2009 (34): 5118 - 5120.

[66] Tang L, Ji R, Cao X, et al. Deep ultraviolet photoluminescence of water - soluble self - passivated graphene quantum dots [J]. ACS Nano, 2012, 6: 5102 - 5110.

[67] Liu X, Li T, Hou Y, et al. Microwave synthesis of carbon dots with multi - response using denatured proteins as carbon source [J]. RSC Advances, 2016, 6 (14): 11711 - 11718.

[68] Wang H - J, Hou W - Y, Yu T - T, et al. Facile microwave synthesis of carbon dots powder with enhanced solid - state fluorescence and its applications in rapid fingerprints detection and white - light - emitting diodes [J]. Dyes and Pigments, 2019,

170: 107623.

[69] Zhai X, Zhang P, Liu C, et al. Highly luminescent carbon nanodots by microwave - assisted pyrolysis [J]. Chemical Communications, 2012, 48 (64): 7955 - 7957.

[70] Chen B, Li F, Li S, et al. Large scale synthesis of photoluminescent carbon nanodots and their application for bioimaging [J]. Nanoscale, 2013, 5 (5): 1967 - 1971.

[71] Ma C B, Zhu Z T, Wang H X, et al. A general solid - state synthesis of chemically - doped fluorescent graphene quantum dots for bioimaging and optoelectronic applications [J]. Nanoscale, 2015, 7 (22): 10162 - 10169.

[72] Martindale B C, Hutton G A, Caputo C A, et al. Solar hydrogen production using carbon quantum dots and a molecular nickel catalyst [J]. Journal of the American Chemical Society, 2015, 137 (18): 6018 - 6025.

[73] Jiang F, Chen D, Li R, et al. Eco - friendly synthesis of size - controllable amine - functionalized graphene quantum dots with antimycoplasma properties [J]. Nanoscale, 2013, 5 (3): 1137 - 1142.

[74] Liu J H, Li R S, Yuan B, et al. Mitochondria - targeting single - layered graphene quantum dots with dual recognition sites for ATP imaging in living cells [J]. Nanoscale, 2018, 10 (36): 17402 - 17408.

[75] Xu X, Tang W, Zhou Y, et al. Steering photoelectrons excited in carbon dots into platinum cluster catalyst for solar - driven hydrogen production [J]. Advanced Science, 2017, 4 (12): 1700273.

[76] Huang X, Zhang F, Zhu L, et al. Effect of injection routes on the biodistribution, clearance, and tumor uptake of carbon dots [J]. ACS Nano, 2013, 7: 5684? 5693.

[77] Yuan F, Wang Z, Li X, et al. Bright multicolor bandgap fluorescent carbon quantum dots for electroluminescent light - emitting diodes [J]. Advanced Materials, 2017, 29 (3): 1604436.

[78] Qu S, Zhou D, Li D, et al. Toward efficient orange emissive carbon nanodots through conjugated sp^2 - domain controlling and surface charges engineering [J]. Advanced Materials, 2016, 28 (18): 3516 - 3521.

[79] Zhan Y, Luo F, Guo L, et al. Preparation of an Efficient Ratiometric Fluorescent Nanoprobe (m - CDs@ $[Ru(bpy)_3]^{2+}$) for Visual and Specific Detection of Hypochlorite on Site and in Living Cells [J]. ACS Sensors, 2017, 2 (11): 1684 - 1691.

[80]Dai Y, Liu Z, Bai Y, et al. A novel highly fluorescent S,N, O co - doped carbon dots for biosensing and bioimaging of copper ions in live cells [J]. RSC Advances, 2018, 8 (73): 42246 - 42252.

[81]Song L, Cui Y, Zhang C, et al. Microwave - assisted facile synthesis of yellow fluorescent carbon dots from o - phenylenediamine for cell imaging and sensitive detection of Fe^{3+} and H_2O_2[J]. RSC Advances, 2016, 6 (21): 17704 - 17712.

[82]Pandit S, Mondal S, De M. Surface engineered amphiphilic carbon dots: solvatochromic behavior and applicability as a molecular probe [J]. Journal of Materials Chemistry B, 2021, 9 (5): 1432 - 1440.

[83]Lu D, Tang Y, Gao J, et al. Concentrated solar irradiation protocols for the efficient synthesis of tri - color emissive carbon dots and photophysical studies [J]. Journal of Materials Chemistry C, 2018, 6 (47): 13013 - 13022.

[84]Vedamalai M, Periasamy A P, Wang C - W, et al. Carbon nanodots prepared from o - phenylenediamine for sensing of Cu^{2+} ions in cells [J]. Nanoscale, 2014, 6 (21): 13119 - 13125.

[85]Liu Q, Niu X, Xie K, et al. Fluorescent Carbon Dots as Nanosensors for Monitoring and Imaging Fe^{3+} and $[HPO_4]^{2-}$ Ions [J]. ACS Applied Nano Materials, 2021, 4 (1): 190 - 197.

[86]Zhao J, Li F, Zhang S, et al. Preparation of N - doped yellow carbon dots and N, P co - doped red carbon dots for bioimaging and photodynamic therapy of tumors [J]. New Journal of Chemistry, 2019, 43 (16): 6332 - 6342.

[87]Shen L, Hou C, Li J, et al. A one - step synthesis of novel high pH - sensitive nitrogen - doped yellow fluorescent carbon dots and their detection application in living cells [J]. Analytical Methods, 2019, 11 (44): 5711 - 5717.

[88]Wang N, Wang M, Yu Y, et al. Label - free fluorescence assay based on near - infrared B,N - doped carbon dots as a fluorescent probe for the detection of sialic acid [J]. New Journal of Chemistry, 2020, 44 (6): 2350 - 2356.

[89]Hu Y, Yang Z, Lu X, et al. Facile synthesis of red dual - emissive carbon dots for ratiometric fluorescence sensing and cellular imaging [J]. Nanoscale, 2020, 12 (9): 5494 - 5500.

[90]Liu Q, Ren B, Xie K, et al. Nitrogen - doped carbon dots for sensitive detection of fer-

ric ions and monohydrogen phosphate by the naked eye and imaging in living cells [J]. Nanoscale Advances, 2021, 3 (3): 805 - 811.

[91] Liu J H, Li D Y, He J H, et al. Polarity - Sensitive Polymer Carbon Dots Prepared at Room - Temperature for Monitoring the Cell Polarity Dynamics during Autophagy [J]. ACS Applied Materials & Interfaces, 2020, 12 (4): 4815 - 4820.

[92] Kalaiyarasan G, Hemlata C, Joseph J. Fluorescence Turn - On, Specific Detection of Cystine in Human Blood Plasma and Urine Samples by Nitrogen - Doped Carbon Quantum Dots [J]. ACS Omega, 2019, 4 (1): 1007 - 1014.

[93] Pan C, Wen Q, Ma L, et al. Green - emissive water - dispersible silicon quantum dots for the fluorescent and colorimetric dual mode sensing of curcumin [J]. Analytical Methods, 2021, 13 (42): 5025 - 5034.

[94] Zhao Y, Geng X, Shi X, et al. A fluorescence - switchable carbon dot for the reversible turn - on sensing of molecular oxygen [J]. Journal of Materials Chemistry C, 2021, 9 (12): 4300 - 4306.

[95] Gao A, Kang Y - F, Yin X - B. Red fluorescence - magnetic resonance dual modality imaging applications of gadolinium containing carbon quantum dots with excitation independent emission [J]. New Journal of Chemistry, 2017, 41 (9): 3422 - 3431.

[96] Phukan K, Sarma R R, Dash S, et al. Carbon dot based nucleus targeted fluorescence imaging and detection of nuclear hydrogen peroxide in living cells [J]. Nanoscale Advances, 2022.

[97] Bai J, Ma Y, Yuan G, et al. Solvent - controlled and solvent - dependent strategies for the synthesis of multicolor carbon dots for pH sensing and cell imaging [J]. Journal of Materials Chemistry C, 2019, 7 (31): 9709 - 9718.

[98] Zhang Z, Pei K, Yang Q, et al. A nanosensor made of sulfur - nitrogen co - doped carbon dots for "off - on" sensing of hypochlorous acid and Zn(ii) and its bioimaging properties [J]. New Journal of Chemistry, 2018, 42 (19): 15895 - 15904.

[99] Lin J - S, Tsai Y - W, Dehvari K, et al. A carbon dot based theranostic platform for dual - modal imaging and free radical scavenging [J]. Nanoscale, 2019, 11 (43): 20917 - 20931.

[100] Shi L, Dong X, Zhang G, et al. Lysosome targeting, Cr(vi) and l - AA sensing, and cell imaging based on N - doped blue - fluorescence carbon dots [J]. Analytical

Methods, 2021, 13 (32): 3561 - 3568.

[101] Xia J, Chen S, Zou G - Y, et al. Synthesis of highly stable red - emissive carbon polymer dots by modulated polymerization: from the mechanism to application in intracellular pH imaging [J]. Nanoscale, 2018, 10 (47): 22484 - 22492.

[102] Pawar S, Kaja S, Nag A. Red - Emitting Carbon Dots as a Dual Sensor for In^{3+} and Pd^{2+} in Water [J]. ACS Omega, 2020, 5 (14): 8362 - 8372.

[103] Li S, Jiang J, Yan Y, et al. Red, green, and blue fluorescent folate - receptor - targeting carbon dots for cervical cancer cellular and tissue imaging [J]. Materials Science and Engineering: C, 2018, 93: 1054 - 1063.

[104] Sato K, Sato R, Iso Y, et al. Surface modification strategy for fluorescence solvatochromism of carbon dots prepared from p - phenylenediamine [J]. Chemical Communications, 2020, 56 (14): 2174 - 2177.

[105] Lu W, Jiao Y, Gao Y, et al. Bright yellow fluorescent carbon dots as a multifunctional sensing platform for the label - free detection of fluoroquinolones and histidine [J]. ACS Applied Materials & Interfaces, 2018, 10 (49): 42915 - 42924.

[106] Li T, E S, Wang J, et al. Regulating the properties of carbon dots via a solvent - involved molecule fusion strategy for improved sensing selectivity [J]. Analytica Chimica Acta, 2019, 1088: 107 - 115.

[107] Sun Z, Zhou W, Luo J, et al. High - efficient and pH - sensitive orange luminescence from silicon - doped carbon dots for information encryption and bio - imaging [J]. Journal of Colloid and Interface Science, 2022, 607: 16 - 23.

[108] Gao D, Zhang Y, Liu A, et al. Photoluminescence - tunable carbon dots from synergy effect of sulfur doping and water engineering [J]. Chemical Engineering Journal, 2020, 388: 124199.

[109] Zhang Q, Wang R, Feng B, et al. Photoluminescence mechanism of carbon dots: triggering high - color - purity red fluorescence emission through edge amino protonation [J]. Nature Communications, 2021, 12 (1): 6856.

[110] Jiao Y, Liu Y, Meng Y, et al. Novel processing for color - tunable luminescence carbon dots and their advantages in biological systems [J]. ACS Sustainable Chemistry & Engineering, 2020, 8 (23): 8585 - 8592.

[111] Hua X - W, Bao Y - W, Zeng J, et al. Nucleolus - targeted red emissive carbon dots

with polarity – sensitive and excitation – independent fluorescence emission: High – resolution cell imaging and in vivo tracking [J]. ACS Applied Materials & Interfaces, 2019, 11 (36): 32647 – 32658.

[112] Song W, Duan W, Liu Y, et al. Ratiometric detection of intracellular lysine and pH with one – pot synthesized dual emissive carbon dots [J]. Analytical Chemistry 2017, 89 (24): 13626 – 13633.

[113] Chen D, Gao H, Chen X, et al. Excitation – independent dual – color carbon dots: surface – state controlling and solid – state lighting [J]. ACS Photonics, 2017, 4 (9): 2352 – 2358.

[114] Shuang E, Mao Q – X, Wang J – H, et al. Carbon dots with tunable dual emissions: from the mechanism to the specific imaging of endoplasmic reticulum polarity [J]. Nanoscale, 2020, 12 (12): 6852 – 6860.

[115] Pang S, Liu S. Dual – emission carbon dots for ratiometric detection of Fe^{3+} ions and acid phosphatase [J]. Analytica Chimica Acta, 2020, 1105: 155 – 161.

[116] Li H, Ye H – G, Cheng R, et al. Red dual – emissive carbon dots for ratiometric sensing of veterinary drugs [J]. Journal of Luminescence, 2021, 236: 118092.

[117] Yang F, Zhou P, Duan C. Solid – phase synthesis of red dual – emissive nitrogen – doped carbon dots for the detection of Cu^{2+} and glutathione [J]. Microchemical Journal, 2021, 169: 106534.

[118] Huang G, Luo X, He X, et al. Dual – emission carbon dots based ratiometric fluorescent sensor with opposite response for detecting copper (II) [J]. Dyes and Pigments, 2021, 196: 109803.

[119] Liu H, He Z, Jiang L P, et al. Microwave – assisted synthesis of wavelength – tunable photoluminescent carbon nanodots and their potential applications [J]. ACS Applied Materials & Interfaces, 2015, 7 (8): 4913 – 4920.

[120] Wang L, Zhu S – J, Wang H – Y, et al. Common origin of green luminescence in carbon nanodots and graphene quantum dots [J]. ACS Nano, 2014, 8: 2541 – 2547.

[121] Zhang Y, Yuan R, He M, et al. Multicolour nitrogen – doped carbon dots: Tunable photoluminescence and sandwich fluorescent glass – based light – emitting diodes [J]. Nanoscale, 2017, 9 (45): 17849 – 17858.

[122] Yan X, Li B, Li L – S. Colloidal graphene quantum dots with well – defined structures

[J]. Accounts of Chemical Research, 2013, 46: 2254 – 2262.

[123] Li Q, Zhang S, Dai L, et al. Nitrogen – doped colloidal graphene quantum dots and their size – dependent electrocatalytic activity for the oxygen reduction reaction [J]. Journal of the American Chemical Society, 2012, 134 (46): 18932 – 18935.

[124] Kim S, Hwang S W, Kim M – K, et al. Anomalous behaviors of visible luminescence from graphene quantum dots: Interplay between size and shape [J]. ACS Nano, 2012, 6: 8203 – 8208.

[125] Zhu S, Song Y, Wang J, et al. Photoluminescence mechanism in graphene quantum dots: Quantum confinement effect and surface/edge state [J]. Nano Today, 2017, 13: 10 – 14.

[126] Essner J B, Kist J A, Polo – Parada L, et al. Artifacts and errors associated with the ubiquitous presence of fluorescent impurities in carbon nanodots [J]. Chemistry of Materials, 2018, 30 (6): 1878 – 1887.

[127] Song Y, Zhu S, Zhang S, et al. Investigation from chemical structure to photoluminescent mechanism: a type of carbon dots from the pyrolysis of citric acid and an amine [J]. Journal of Materials Chemistry C, 2015, 3 (23): 5976 – 5984.

[128] Schneider J, Reckmeier C J, Xiong Y, et al. Molecular fluorescence in citric acid – based carbon dots [J]. The Journal of Physical Chemistry C, 2017, 121 (3): 2014 – 2022.

[129] Righetto M, Privitera A, Fortunati I, et al. Spectroscopic insights into carbon dot systems [J]. The Journal of Physical Chemistry Letters, 2017, 8 (10): 2236 – 2242.

[130] Cao L, Wang X, Meziani M J, et al. Carbon dots for multiphoton bioimaging [J]. Journal of the American Chemical Society, 2007, 129: 11318 – 11319.

[131] Jia X, Li J, Wang E. One – pot green synthesis of optically pH – sensitive carbon dots with upconversion luminescence [J]. Nanoscale, 2012, 4 (18): 5572 – 5575.

[132] Wang C, Wu X, Li X, et al. Upconversion fluorescent carbon nanodots enriched with nitrogen for light harvesting [J]. Journal of Materials Chemistry, 2012, 22 (31): 15522.

[133] Li H, He X, Liu Y, et al. One – step ultrasonic synthesis of water – soluble carbon nanoparticles with excellent photoluminescent properties [J]. Carbon, 2011, 49 (2): 605 – 609.

[134] Liu Q, Guo B, Rao Z, et al. Strong two - photon - induced fluorescence from photostable, biocompatible nitrogen - doped graphene quantum dots for cellular and deep - tissue imaging [J]. Nano Letters, 2013, 13 (6): 2436 - 2441.

[135] Shen J, Zhu Y, Chen C, et al. Facile preparation and upconversion luminescence of graphene quantum dots [J]. Chemical Communications, 2011, 47: 2580 - 2582.

[136] Gan N, Shi H, An Z, et al. Recent advances in polymer - based metal - free room - temperature phosphorescent materials [J]. Advanced Functional Materials, 2018, 28 (51): 1802657.

[137] Jiang K, Wang Y, Li Z, et al. Afterglow of carbon dots: Mechanism, strategy and applications [J]. Materials Chemistry Frontiers, 2020, 4 (2): 386 - 399.

[138] Su Q, Lu C, Yang X. Efficient room temperature phosphorescence carbon dots: Information encryption and dual - channel pH sensing [J]. Carbon, 2019, 152: 609 - 615.

[139] Jiang K, Zhang L, Lu J, et al. Triple - mode emission of carbon dots: Applications for advanced anti - counterfeiting [J]. Angewandte Chemie International Edition, 2016, 55 (25): 7231 - 7235.

[140] Miao S, Liang K, Zhu J, et al. Hetero - atom - doped carbon dots: Doping strategies, properties and applications [J]. Nano Today, 2020, 33: 100879.

[141] Lv J J, Zhao J, Fang H, et al. Incorporating nitrogen - doped graphene quantum dots and Ni3 S2 nanosheets: A synergistic electrocatalyst with highly enhanced activity for overall water splitting [J]. Small, 2017, 13 (24): 1700264.

[142] Li Q, Zhou M, Yang Q, et al. Efficient room - temperature phosphorescence from nitrogen - doped carbon dots in composite matrices [J]. Chemistry of Materials, 2016, 28 (22): 8221 - 8227.

[143] Zhang Y - Q, Ma D - K, Zhuang Y, et al. One - pot synthesis of N - doped carbon dots with tunable luminescence properties [J]. Journal of Materials Chemistry, 2012, 22 (33): 16714 - 16718.

[144] Peng H, Li Y, Jiang C, et al. Tuning the properties of luminescent nitrogen - doped carbon dots by reaction precursors [J]. Carbon, 2016, 100: 386 - 394.

[145] Liu Y, Jiang L, Li B, et al. Nitrogen doped carbon dots: mechanism investigation and their application for label free CA125 analysis [J]. Journal of Materials Chemistry B,

2019, 7 (19): 3053 -3058.

[146] Lu Y - C, Chen J, Wang A - J, et al. Facile synthesis of oxygen and sulfur co - doped graphitic carbon nitride fluorescent quantum dots and their application for mercury(II) detection and bioimaging [J]. Journal of Materials Chemistry C, 2015, 3 (1): 73 -78.

[147] Yang G, Wan X, Su Y, et al. Acidophilic S - doped carbon quantum dots derived from cellulose fibers and their fluorescence sensing performance for metal ions in an extremely strong acid environment [J]. Journal of Materials Chemistry A, 2016, 4 (33): 12841 -12849.

[148] Chandra S, Patra P, Pathan S H, et al. Luminescent S - doped carbon dots: An emergent architecture for multimodal applications [J]. Journal of Materials Chemistry B, 2013, 1 (18): 2375 -2382.

[149] Lu W, Gong X, Nan M, et al. Comparative study for N and S doped carbon dots: Synthesis, characterization and applications for Fe^{3+} probe and cellular imaging [J]. Analytica Chimica Acta, 2015, 898: 116 -127.

[150] Han Y, Tang D, Yang Y, et al. Non - metal single/dual doped carbon quantum dots: A general flame synthetic method and electro - catalytic properties [J]. Nanoscale, 2015, 7 (14): 5955 -5962.

[151] Hua J, Jiao Y, Wang M, et al. Determination of norfloxacin or ciprofloxacin by carbon dots fluorescence enhancement using magnetic nanoparticles as adsorbent [J]. Microchim Acta, 2018, 185 (2): 137.

[152] Shan X, Chai L, Ma J, et al. B - doped carbon quantum dots as a sensitive fluorescence probe for hydrogen peroxide and glucose detection [J]. Analyst, 2014, 139 (10): 2322 -2325.

[153] Bourlinos A B, Trivizas G, Karakassides M A, et al. Green and simple route toward boron doped carbon dots with significantly enhanced non - linear optical properties [J]. Carbon, 2015, 83: 173 -179.

[154] Shen C, Wang J, Cao Y, et al. Facile access to B - doped solid - state fluorescent carbon dots toward light emitting devices and cell imaging agents [J]. Journal of Materials Chemistry C, 2015, 3 (26): 6668 -6675.

[155] Niu X, Song T, Xiong H. Large scale synthesis of red emissive carbon dots powder by

solid state reaction for fingerprint identification [J]. Chinese Chemical Letters, 2021, 32 (6): 1953 -1956.

[156] Chandra S, Das P, Bag S, et al. Synthesis, functionalization and bioimaging applications of highly fluorescent carbon nanoparticles [J]. Nanoscale, 2011, 3 (4): 1533 -1540.

[157] Zhou J, Shan X, Ma J, et al. Facile synthesis of P - doped carbon quantum dots with highly efficient photoluminescence [J]. RSC Advances, 2014, 4 (11): 5465 -5468.

[158] Xu Q, Liu Y, Su R, et al. Highly fluorescent Zn - doped carbon dots as Fenton reaction - based bio - sensors: An integrative experimental - theoretical consideration [J]. Nanoscale, 2016, 8 (41): 17919 -17927.

[159] Bourlinos A B, Bakandritsos A, Kouloumpis A, et al. Gd(III) - doped carbon dots as a dual fluorescent - MRI probe [J]. Journal of Materials Chemistry, 2012, 22 (44): 23327.

[160] Dong Y, Pang H, Yang H B, et al. Carbon - based dots co - doped with nitrogen and sulfur for high quantum yield and excitation - independent emission [J]. Angewandte Chemie International Edition, 2013, 52 (30): 7800 -7804.

[161] Shi B, Su Y, Zhang L, et al. Nitrogen and phosphorus co - doped carbon nanodots as a novel fluorescent probe for highly sensitive detection of Fe^{3+} in human serum and living cells [J]. ACS Applied Materials & Interfaces, 2016, 8 (17): 10717 -10725.

[162] Tan L, Huang G, Liu T, et al. Synthesis of highly bright oil - soluble carbon quantum dots by hot - injection method with N and B co - doping [J]. Journal of Nanoscience and Nanotechnology, 2016, 16 (3): 2652 -2657.

[163] Zhou J, Zhou H, Tang J, et al. Carbon dots doped with heteroatoms for fluorescent bioimaging: A review [J]. Microchimica Acta, 2016, 184 (2): 343 -368.

[164] Yang M, Feng T, Chen Y, et al. Ionic - state cobalt and iron co - doped carbon dots with superior electrocatalytic activity for the oxygen evolution reaction [J]. ChemElectroChem, 2019, 6 (7): 2088 -2094.

[165] Wang X, Ma Y, Wu Q, et al. Ultra - bright and stable pure blue light - emitting diode from O, N co - doped carbon dots [J]. Laser & Photonics Reviews, 2021.

[166] Jiang L, Ding H, Lu S, et al. Photoactivated fluorescence enhancement in F, N - doped carbon dots with piezochromic behavior [J]. Angewandte Chemie International

Edition, 2020, 59: 9986 - 9991.

[167] Liu T, Li N, Dong J X, et al. Fluorescence detection of mercury ions and cysteine based on magnesium and nitrogen co - doped carbon quantum dots and IMPLICATION logic gate operation [J]. Sensors and Actuators B: Chemical, 2016, 231: 147 - 153.

[168] Huang S, Yang E, Yao J, et al. Nitrogen, phosphorus and sulfur tri - doped carbon dots are specific and sensitive fluorescent probes for determination of chromium(VI) in water samples and in living cells [J]. Microchimica Acta, 2019, 186 (12): 851.

[169] Liu Y, Zhang T, Wang R, et al. A facile and universal strategy for preparation of long wavelength emission carbon dots [J]. Dalton Transactions, 2017, 46 (48): 16905 - 16910.

[170] Zhan J, Geng B, Wu K, et al. A solvent - engineered molecule fusion strategy for rational synthesis of carbon quantum dots with multicolor bandgap fluorescence [J]. Carbon, 2018, 130: 153 - 163.

[171] Ding H, Wei J - S, Zhang P, et al. Solvent - controlled synthesis of highly luminescent carbon dots with a wide color gamut and narrowed emission peak widths [J]. Small, 2018, 14 (22): 1800612.

[172] Wang X, Cao L, Yang S T, et al. Bandgap - like strong fluorescence in functionalized carbon nanoparticles [J]. Angewandte Chemie International Edition, 2010, 49 (31): 5310 - 5314.

[173] Tetsuka H, Asahi R, Nagoya A, et al. Optically tunable amino - functionalized graphene quantum dots [J]. Advanced Materials, 2012, 24 (39): 5333 - 5338.

[174] Tetsuka H, Nagoya A, Fukusumi T, et al. Molecularly designed, nitrogen - functionalized graphene quantum dots for optoelectronic devices [J]. Advanced Materials, 2016, 28 (23): 4632 - 4638.

[175] Li M, Hu C, Yu C, et al. Organic amine - grafted carbon quantum dots with tailored surface and enhanced photoluminescence properties [J]. Carbon, 2015, 91: 291 - 297.

[176] Gupta V, Chaudhary N, Srivastava R, et al. Luminscent graphene quantum dots for organic photovoltaic devices [J]. Journal of the American Chemical Society, 2011, 133 (26): 9960 - 9963.

[177] Dhenadhayalan N, Lin K C, Saleh T A. Recent advances in functionalized carbon dots

toward the design of efficient materials for sensing and catalysis applications [J]. Small, 2020, 16 (1): 1905767.

[178] Gao W, Song H, Wang X, et al. Carbon dots with red emission for sensing of Pt^{2+}, Au^{3+}, and Pd^{2+} and their bioapplications in vitro and in vivo [J]. ACS Applied Materials & Interfaces, 2018, 10 (1): 1147 - 1154.

[179] Zhang M, Wang W, Yuan P, et al. Synthesis of lanthanum doped carbon dots for detection of mercury ion, multi - color imaging of cells and tissue, and bacteriostasis [J]. Chemical Engineering Journal, 2017, 330: 1137 - 1147.

[180] Lesani P, Singh G, Viray C M, et al. Two - photon dual - emissive carbon dot - based probe: Deep - tissue imaging and ultrasensitive sensing of intracellular ferric ions [J]. ACS Applied Materials & Interfaces, 2020, 12 (16): 18395 - 18406.

[181] Yang M, Tang Q, Meng Y, et al. Reversible "off - on" fluorescence of Zn^{2+} - passivated carbon dots: Mechanism and potential for the detection of EDTA and Zn^{2+} [J]. Langmuir, 2018, 34 (26): 7767 - 7775.

[182] Yang P, Zhu Z, Zhang T, et al. Orange - emissive carbon quantum dots: Toward application in wound pH monitoring based on colorimetric and fluorescent changing [J]. Small, 2019, 15 (44): 1902823.

[183] Gao Y, Jiao Y, Lu W, et al. Carbon dots with red emission as a fluorescent and colorimeteric dual - readout probe for the detection of chromium(Ⅵ) and cysteine and its logic gate operation [J]. Journal of Materials Chemistry B, 2018, 6 (38): 6099 - 6107.

[184] Jiao Y, Gao Y, Meng Y, et al. One - step synthesis of label - free ratiometric fluorescence carbon dots for the detection of silver ions and glutathione and cellular imaging applications [J]. ACS Applied Materials & Interfaces, 2019, 11 (18): 16822 - 16829.

[185] Yao W, Wu N, Lin Z, et al. Fluorescent turn - off competitive immunoassay for biotin based on hydrothermally synthesized carbon dots [J]. Microchimica Acta, 2017, 184 (3): 907 - 914.

[186] Sun Z, Chen Z, Luo J, et al. A yellow - emitting nitrogen - doped carbon dots for sensing of vitamin B12 and their cell - imaging [J]. Dyes and Pigments, 2020, 176: 108227.

[187] Hu G, Ge L, Li Y, et al. Carbon dots derived from flax straw for highly sensitive and selective detections of cobalt, chromium, and ascorbic acid [J]. Journal of Colloid and Interface Science, 2020, 579: 96 - 108.

[188] Liu H, Sun Y, Li Z, et al. Lysosome - targeted carbon dots for ratiometric imaging of formaldehyde in living cells [J]. Nanoscale, 2019, 11 (17): 8458 - 8463.

[189] Gu J, Li X, Zhou Z, et al. 2D MnO_2 nanosheets generated signal transduction with 0D carbon quantum dots: Synthesis strategy, dual - mode behavior and glucose detection [J]. Nanoscale, 2019, 11 (27): 13058 - 13068.

[190] Gong P, Sun L, Wang F, et al. Highly fluorescent N - doped carbon dots with two - photon emission for ultrasensitive detection of tumor marker and visual monitor anticancer drug loading and delivery [J]. Chemical Engineering Journal, 2019, 356: 994 - 1002.

[191] Yang D, Guo Z, Wang J, et al. Carbon nanodot - based fluorescent method for virus DNA analysis with isothermal strand displacement amplification [J]. Particle & Particle Systems Characterization, 2019, 36 (10): 1900273.

[192] Gao G, Jiang Y W, Yang J, et al. Mitochondria - targetable carbon quantum dots for differentiating cancerous cells from normal cells [J]. Nanoscale, 2017, 9 (46): 18368 - 18378.

[193] Cheng W, Xu J, Guo Z, et al. Hydrothermal synthesis of N,S co - doped carbon nanodots for highly selective detection of living cancer cells [J]. Journal of Materials Chemistry B, 2018, 6 (36): 5775 - 5780.

[194] Zhao X, Tang Q, Zhu S, et al. Controllable acidophilic dual - emission fluorescent carbonized polymer dots for selective imaging of bacteria [J]. Nanoscale, 2019, 11 (19): 9526 - 9532.

[195] Hoang V C, Dave K, Gomes V G. Carbon quantum dot - based composites for energy storage and electrocatalysis: Mechanism, applications and future prospects [J]. Nano Energy, 2019, 66: 104093.

[196] Han M, Zhu S, Lu S, et al. Recent progress on the photocatalysis of carbon dots: Classification, mechanism and applications [J]. Nano Today, 2018, 19: 201 - 218.

[197] Yang M, Feng T, Chen Y, et al. Synchronously integration of Co, Fe dual - metal doping in Ru@ C and CDs for boosted water splitting performances in alkaline media

[J]. Applied Catalysis B: Environmental, 2020, 267: 118657.

[198] Cailotto S, Negrato M, Daniele S, et al. Carbon dots as photocatalysts for organic synthesis: Metal – free methylene – oxygen – bond photocleavage [J]. Green Chemistry, 2020, 22 (4): 1145 – 1149.

[199] Han M, Lu S, Qi F, et al. Carbon dots – implanted graphitic carbon nitride nanosheets for photocatalysis: Simultaneously manipulating carrier transport in inter – and intralayers [J]. Solar RRL, 2020, 4 (4): 1900517.

[200] Xu L, Bai X, Guo L, et al. Facial fabrication of carbon quantum dots (CDs) – modified N – TiO_2 – x nanocomposite for the efficient photoreduction of Cr(VI) under visible light [J]. Chemical Engineering Journal, 2019, 357: 473 – 486.

[201] Wang S, Li L, Zhu Z, et al. Remarkable improvement in photocatalytic performance for tannery wastewater processing via SnS_2 modified with N – doped carbon quantum dots: Synthesis, characterization, and 4 – nitrophenol – aided Cr(VI) photoreduction [J]. Small, 2019, 15 (29): 1804515.

[202] Huang S, Jiang S, Pang H, et al. Dual functional nanocomposites of magnetic MnFe2O4 and fluorescent carbon dots for efficient U(VI) removal [J]. Chemical Engineering Journal, 2019, 368: 941 – 950.

[203] Luo H, Liu Y, Dimitrov S D, et al. Pt single – atoms supported on nitrogen – doped carbon dots for highly efficient photocatalytic hydrogen generation [J]. Journal of Materials Chemistry A, 2020, 8 (29): 14690 – 14696.

[204] Yan Y, Chen J, Li N, et al. Systematic bandgap engineering of graphene quantum dots and applications for photocatalytic water splitting and CO_2 reduction [J]. ACS Nano, 2018, 12 (4): 3523 – 3532.

[205] Li M, Wang M, Zhu L, et al. Facile microwave assisted synthesis of N – rich carbon quantum dots/dual – phase TiO_2 heterostructured nanocomposites with high activity in CO_2 photoreduction [J]. Applied Catalysis B: Environmental, 2018, 231: 269 – 276.

[206] Sarma D, Majumdar B, Sarma T K. Visible – light induced enhancement in the multi – catalytic activity of sulfated carbon dots for aerobic carbon – carbon bond formation [J]. Green Chemistry, 2019, 21 (24): 6717 – 6726.

[207] Bhattacharyya S, Ehrat F, Urban P, et al. Effect of nitrogen atom positioning on the

trade – off between emissive and photocatalytic properties of carbon dots [J]. Nat Commun, 2017, 8 (1): 1401.

[208] Guo S, Zhao S, Wu X, et al. A Co_3O_4 – CDots – C_3N_4 three component electrocatalyst design concept for efficient and tunable CO_2 reduction to syngas [J]. Nat Commun, 2017, 8 (1): 1828.

[209] Liu C, Fu Y, Xia Y, et al. Cascaded photo – potential in a carbon dot – hematite system driving overall water splitting under visible light [J]. Nanoscale, 2018, 10 (5): 2454 – 2460.

[210] Wu Y, Zhang H, Pan A, et al. White – light – emitting melamine – formaldehyde microspheres through polymer – mediated aggregation and encapsulation of graphene quantum dots [J]. Advanced Science, 2019, 6 (2): 1801432.

[211] Wang C, Hu T, Chen Y, et al. Polymer – assisted self – assembly of multicolor carbon dots as solid – state phosphors for fabrication of warm, high – quality, and tem perature – responsive white – light – emitting devices [J]. ACS Applied Materials & Interfaces, 2019, 11 (25): 22332 – 22338.

[212] Yan F, Jiang Y, Sun X, et al. Multicolor carbon dots with concentration – tunable fluorescence and solvent – affected aggregation states for white light – emitting diodes [J]. Nano Research, 2019, 13 (1): 52 – 60.

[213] Zhan Y, Shang B, Chen M, et al. One – step synthesis of silica – coated carbon dots with controllable solid – state fluorescence for white light – emitting diodes [J]. Small, 2019, 15 (24): 1901161.

[214] Zhou D, Zhai Y, Qu S, et al. Electrostatic assembly guided synthesis of highly luminescent carbon – nanodots@ $BaSO_4$ hybrid phosphors with improved stability [J]. Small, 2017, 13 (6): 1602055.

[215] Zhou D, Jing P, Wang Y, et al. Carbon dots produced via space – confined vacuum heating: Maintaining efficient luminescence in both dispersed and aggregated states [J]. Nanoscale Horizons, 2019, 4 (2): 388 – 395.

[216] He J, He Y, Chen Y, et al. Solid – state carbon dots with red fluorescence and efficient construction of dual – fluorescence morphologies [J]. Small, 2017, 13 (26): 1700075.

[217] Wang F, Chen Y – H, Liu C – Y, et al. White light – emitting devices based on car-

bon dots' electroluminescence [J]. Chemical Communications, 2011, 47: 3502 – 3504.

[218] Yuan F, Yuan T, Sui L, et al. Engineering triangular carbon quantum dots with unprecedented narrow bandwidth emission for multicolored LEDs [J]. Nature Communications, 2018, 9 (1): 2249.

[219] Yuan F, Wang Y – K, Sharma G, et al. Bright high – colour – purity deep – blue carbon dot light – emitting diodes via efficient edge amination [J]. Nature Photonics, 2019, 14 (3): 171 – 176.

[220] Zhang Y, Foster C W, Banks C E, et al. Graphene – rich wrapped petal – like rutile TiO_2 tuned by carbon dots for high – performance sodium storage [J]. Advanced Materials, 2016, 28 (42): 9391 – 9399.

[221] Zhang E, Jia X, Wang B, et al. Carbon dots@ rGO paper as freestanding and flexible potassium – ion batteries anode [J]. Advanced Science, 2020, 7 (15): 2000470.

[222] Zhang Q, Sun C, Fan L, et al. Iron fluoride vertical nanosheets array modified with graphene quantum dots as long – life cathode for lithium ion batteries [J]. Chemical Engineering Journal, 2019, 371: 245 – 251.

[223] Yin X, Chen H, Zhi C, et al. Functionalized graphene quantum dot modification of yolk – shell NiO microspheres for superior lithium storage [J]. Small, 2018, 14 (22): 1800589.

[224] Ma C, Dai K, Hou H, et al. High ion – conducting solid – state composite electrolytes with carbon quantum dot nanofillers [J]. Advanced Science, 2018, 5 (5): 1700996.

[225] Zheng C, Luo N, Huang S, et al. Nanocomposite of Mo2N quantum dots@ MoO_3@ nitrogen – doped carbon as a high – performance anode for lithium – ion batteries [J]. ACS Sustainable Chemistry & Engineering, 2019, 7 (12): 10198 – 10206.

[226] Hu Y, Chen W, Lei T, et al. Carbon quantum dots – modified interfacial interactions and ion conductivity for enhanced high current density performance in lithium – sulfur batteries [J]. Advanced Energy Materials, 2019, 9 (7): 1802955.

[227] E S, Mao Q – X, Yuan X – L, et al. Targeted imaging of the lysosome and endoplasmic reticulum and their pH monitoring with surface regulated carbon dots [J]. Nanoscale, 2018, 10: 12788 – 12796.

[228] Sun S, Zhang L, Jiang K, et al. Toward high – efficient red emissive carbon dots: Facile preparation, unique properties, and applications as multifunctional theranostic

agents [J]. Chemistry of Materials, 2016, 28 (23): 8659 -8668.

[229] Liu J, Lu S, Tang Q, et al. One - step hydrothermal synthesis of photoluminescent carbon nanodots with selective antibacterial activity against Porphyromonas gingivalis [J]. Nanoscale, 2017, 9 (21): 7135 -7142.

[230] Bouzas - Ramos D, Cigales Canga J, Mayo J C, et al. Carbon quantum dots codoped with nitrogen and lanthanides for multimodal imaging [J]. Advanced Functional Materials, 2019, 29 (38): 1903884.

[231] Liu Y, Liu J, Zhang J, et al. A brand - new generation of fluorescent nano - neural tracers: Biotinylated dextran amine conjugated carbonized polymer dots [J]. Biomaterials Science, 2019, 7 (4): 1574 -1583.

[232] Song Y, Li H, Lu F, et al. Fluorescent carbon dots with highly negative charges as a sensitive probe for real - time monitoring of bacterial viability [J]. Journal of Materials Chemistry B, 2017, 5 (30): 6008 -6015.

[233] Lu F, Song Y, Huang H, et al. Fluorescent carbon dots with tunable negative charges for bio - imaging in bacterial viability assessment [J]. Carbon, 2017, 120: 95 -102.

[234] Lin F, Li C, Dong L, et al. Imaging biofilm - encased microorganisms using carbon dots derived from L. plantarum [J]. Nanoscale, 2017, 9 (26): 9056 -9064.

[235] Yang J, Gao G, Zhang X, et al. One - step synthesis of carbon dots with bacterial contact - enhanced fluorescence emission: Fast Gram - type identification and selective Gram - positive bacterial inactivation [J]. Carbon, 2019, 146: 827 -839.

[236] Li H, Huang J, Song Y, et al. Degradable carbon dots with broad - spectrum antibacterial activity [J]. ACS Applied Materials & Interfaces, 2018, 10 (32): 26936 -26946.

[237] Wang S, Zhang Y, Zhuo P, et al. Identification of eight pathogenic microorganisms by single concentration - dependent multicolor carbon dots [J]. Journal of Materials Chemistry B, 2020, 8 (27): 5877 -5882.

[238] Li W, Zhang H, Zheng Y, et al. Multifunctional carbon dots for highly luminescent orange - emissive cellulose based composite phosphor construction and plant tissue imaging [J]. Nanoscale, 2017, 9 (35): 12976 -12983.

[239] Ji Z, Arvapalli D M, Zhang W, et al. Nitrogen and sulfur co - doped carbon nanodots in living EA. hy926 and A549 cells: oxidative stress effect and mitochondria targeting [J]. Journal of Materials Science, 2020, 55 (14): 6093 -6104.

[240] Geng X, Sun Y, Li Z, et al. Retrosynthesis of tunable fluorescent carbon dots for precise long – term mitochondrial tracking [J]. Small, 2019, 15 (48): e1901517.

[241] Liu Y, Liu J, Zhang J, et al. Noninvasive brain tumor imaging using red emissive carbonized polymer dots across the blood – brain barrier [J]. ACS Omega, 2018, 3 (7): 7888 – 7896.

[242] Wang L, Wu B, Li W, et al. Industrial production of ultra – stable sulfonated graphene quantum dots for Golgi apparatus imaging [J]. Journal of Materials Chemistry B, 2017, 5 (27): 5355 – 5361.

[243] Li R S, Gao P F, Zhang H Z, et al. Chiral nanoprobes for targeting and long – term imaging of the Golgi apparatus [J]. Chemical Science, 2017, 8 (10): 6829 – 6835.

[244] Hua X W, Bao Y W, Wu F G. Fluorescent carbon quantum dots with intrinsic nucleolus – targeting capability for nucleolus imaging and enhanced cytosolic and nuclear drug delivery [J]. ACS Applied Materials & Interfaces, 2018, 10 (13): 10664 – 10677.

[245] Liu H, Yang J, Li Z, et al. Hydrogen – bond – induced emission of carbon dots for wash – free nucleus imaging [J]. Analytical Chemistry, 2019, 91 (14): 9259 – 9265.

[246] Li J, Yang S, Deng Y, et al. Emancipating target – functionalized carbon dots from autophagy vesicles for a novel visualized tumor therapy [J]. Advanced Functional Materials, 2018, 28 (30): 1800881.

[247] Sung S Y, Su Y L, Cheng W, et al. Graphene quantum dots – mediated theranostic penetrative delivery of drug andphotolytics in deep tumors by targeted biomimetic nanosponges [J]. Nano Letters, 2019, 19 (1): 69 – 81.

[248] Li S, Su W, Wu H, et al. Targeted tumour theranostics in mice via carbon quantum dots structurally mimicking large amino acids [J]. Nature Biomedical Engineering, 2020, 4 (7): 704 – 716.

[249] Gao P, Liu S, Su Y, et al. Fluorine – doped carbon dots with intrinsic nucleus – targeting ability for drug and dye delivery [J]. Bioconjugate Chemistry, 2020, 31 (3): 646 – 655.

[250] Scialabba C, Sciortino A, Messina F, et al. Highly homogeneous biotinylated carbon nanodots: Red – emitting nanoheaters as theranostic agents toward precision cancer medicine [J]. ACS Applied Materials & Interfaces, 2019, 11 (22): 19854 – 19866.

[251] Ghosh S, Ghosal K, Mohammad S A, et al. Dendrimer functionalized carbon quantum

dot for selective detection of breast cancer and gene therapy [J]. Chemical Engineering Journal, 2019, 373: 468 – 484.

[252] Han J, Na K. Transfection of the TRAIL gene into human mesenchymal stem cells using biocompatible polyethyleneimine carbon dots for cancer gene therapy [J]. Journal of Industrial and Engineering Chemistry, 2019, 80: 722 – 728.

第2章

反应溶剂参与的分子融合法调控碳点表面态用于改善其检测选择性

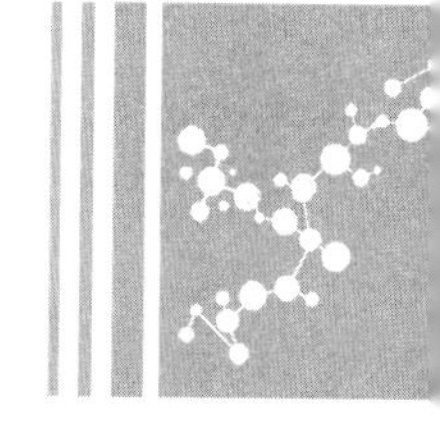

2.1 引言

CDs 作为荧光纳米材料家族的新成员，由于其杰出的光学性能、优异的水分散性和良好的生物相容性等而备受关注，也因此在传感、生物成像、药物运输和光催化领域得到应用[1-10]。溶剂热法是 CDs 制备中最常用的方法，基于溶剂热法通过选择合适的前驱体并控制反应温度、时间等条件，可以轻松调控 CDs 的物理化学性质，以满足实际需求。Zhang 等人通过调节柠檬酸铵和乙二胺四乙酸比例制备得到了具有连续全色发射的 CDs，同时 CDs 的量子产率可以达到 32.8%[11]。Liu 等人的研究证明，由特定蛋白质所制备得到的 CDs 荧光会受某些金属离子的影响发生淬灭，即蛋白酶和脂肪酶衍生的 CDs 分别显示出对 Cu^{2+} 和 Ni^{2+} 的检测选择性。这表明，通过筛选蛋白质来源以制备适当 CDs 可以实现一种非常简单且可行的金属离子检测系统[12]。除了碳源的选择对 CDs 的性质有影响外，CDs 制备过程中调控反应温度也会影响 CDs 的性能。例如：Lu 等发现制备 CDs 时的温度会影响 CDs 的结构进而影响 CDs 的荧光发射。在低温度下制备的 CDs 主要包含大分子和交联的聚合物链而不是碳化的结构。此时 CDs 荧光发射机理来自于分子发射或交联增强发射。随着制备温度的升高，碳核在 CDs 中占主导地位，使其具有特征性的蓝色发射[13]。此外，CDs 的性能还可以通过选择合适的反应溶剂来调节。例如，Zhang 等以对苯二胺作为碳源，研究了在相同的制备条件下不同反应溶剂对得到 CDs 性能的影响。研究发现，反应溶剂的极性的改变会影响 CDs 的荧光发射颜色[14]。Zhan 课题组通过改变反应溶剂的组成，系统地控制了 CDs 的带隙，得到了不同荧光发射的 CDs[15]。由此可见，在 CDs 制备过程中任何一个条件的改变都会对 CDs 的性能有所影响，因此研究外界因素（例如：碳源、反应温度、反应溶剂以及反应时间）改变所引起的 CDs 性质变化的规律，对于全面了解 CDs 的形成机理以及拓宽 CDs 在多个领域的应用是非常必要的。

随着社会进步和工业化进程的快速发展，诸多的重金属污染物被违规排放到了生态系统中，使环境污染形势严峻且不容乐观[16]。这些重金属因其高活性、高毒性、长持久性、生物难分解性、易富集和对生物体作用受到了广泛重视[17,18]。难降解的重金属以不同的状态或者形态在进入到自然环境或者生态系统后会不断被富集积累，从而影响到生态系统的平衡，并且重金属在进入到水体环境后会在底泥中沉积下来，被水生生物摄取，继而通过食物链层层积累，最终人类食用水产品导致重金属进入人体，对人体造成不可估量的健康危害[19]。因此，重金属离子的检测与人们的安全生产生活有着紧密的联系。众所周知,银在电器科学、摄影和医药工业方面有着举足轻重的作用[20]。但是,由于违规排放导致其在环境中大量存在,它的潜在毒性也受到了人们的广泛关注。研究发现,Ag^+可以与多种代谢物和酶相互作用,进而导致这些物质丧失功能,如含有巯基的酶的活性[21]。半胱氨酸(Cys)在20多种氨基酸中是唯一含有巯基(-SH)的氨基酸,巯基具有还原性,使得Cys具有很多功能[22]。现今Cys已在很多领域广泛应用,可用于芳香族酸和丙烯腈中毒后的解毒;制造果汁时可以防止果汁变色和维生素C被氧化;制作面包时可以促进发酵和防止老化;可以预防人体被放射线损伤;可以加入护肤膏霜和美容水等化妆品中;可以用于治疗支气管炎[22]。Cys广泛参与到生物体的重要生理活动中。生长缓慢、头发褪色、水肿、肝损伤以及皮肤受损都与Cys的不正常水平有关[23]。

综上所述,开发一种对Ag^+和Cys具有选择性和灵敏性的分析检测方法是十分必要的。实际上,目前已经有一些基于金属离子对CDs的荧光淬灭作用,将CDs作为荧光探针用于金属离子定量分析的报道[24]。CDs在这些检测中表现出了极强的选择性和灵敏性。在本章的研究中,采用反应溶剂参与的分子融合法以o-PD为碳源,选择不同的反应溶剂制备了一系列o-PD衍生CDs。深入研究了反应溶剂种类对o-PD衍生CDs的表面态和检测性能的影响,其中以甲酰胺为反应溶剂制备的FCDs为主要的研究对象。研究发现,在FCDs的结构中不仅具有从碳源和反应溶剂保留的官能团,而且还存在大量的C=N基团,这是由于o-PD的$-NH_2$与甲酰胺的C=O之间的席夫碱反应所致。由于这些官能团的存在,导致FCDs在Ag^+检测中表现出更好的性能。值得注意的是,与先前报道的用于Ag^+检测的CDs相比,制备的FCDs对其他金属离子的耐受性显著提高。基于此,FCDs被开发为荧光探针以荧光"关闭"方式检测Ag^+。同时,由于Ag^+与Cys之间的相互作用,FCDs/Ag^+体系可以通过荧光"开启"方式用于Cys的定量分析,实现了在FCDs探针平台上的以荧光"开关"方式同时检测Ag^+与Cys。此外,制备的FCDs还显示出了良好的生物相容性和较低的细胞毒性,并且可以用作细胞中Ag^+和Cys成像的有效荧光纳米探针,同时FCDs的长波波长荧光发射还可以有效防止动物组织自身荧光的干扰。

2.2 实验部分

2.2.1 实验仪器

本章所使用的仪器品牌和型号如下:

MPC－5V316 型医用冷藏保存箱（中国安徽中科都菱有限公司）；

KQ－100B/800KDE 型超声波清洗器（中国昆山市超声仪器有限公司）；

BSA22AS 单盘型分析电子天平（中国北京赛多利斯仪器有限公司）；

TG16－WS 台式高速离心机（中国湘仪实验室仪器开发有限公司）；

2XZ－2 型真空泵（中国临海市谭式真空设备有限公司）；

PB－10 标准型 pH 计（中国北京赛多利斯仪器有限公司）；

Lambda Bio20 紫外可见光分光光度计（美国珀金埃尔默仪器有限公司）；

F－7000 荧光分光光度计（日本日立公司）；

JEM－2100 透射电子显微镜（日本电子株式会社）；

Bruker Dimension icon 原子力显微镜（德国布鲁克公司）；

DZF－6020 型真空干燥箱（中国上海精宏实验设备有限公司）；

One Spectra 红外光谱仪（美国珀金埃尔默仪器有限公司）；

DHG－9037A 电热恒温干燥箱（中国上海精宏实验设备有限公司）；

702 型超低温冰箱（美国赛默飞世尔公司）；

EscaLab 250Xi X 射线光电子能谱分析仪（美国赛默飞世尔公司）；

SW－CJ－2FD 型超净台（中国江苏苏州安泰技术有限公司）；

HERA Cell 150 型细胞培养箱（美国赛默飞世尔公司）；

HVE－50 型高压灭菌箱（日本平山公司）；

Synergy H1 型酶标仪（美国 BioTek 公司）；

YE5A44 型手动可调式移液器（中国上海大龙医疗设备有限公司）。

2.2.2　实验试剂

本章所使用化学试剂和品牌如下：

O－PD、Cys、葡萄糖、抗坏血酸（AA）、硫胺素、谷胱甘肽（GSH）以及本章使用的其他氨基酸购自阿拉丁化学有限公司（中国上海）。甲酰胺、乙醇、N，N－二甲基甲酰胺（DMF）、二甲基亚砜（DMSO）、硝酸银（$AgNO_3$）以及本章使用的其他无机盐购自国药集团化学试剂有限公司（中国上海）。胎牛血清（FBS）、高糖培养基（DMEM）、胰酶（0.25%）、青霉素、链霉素购自美国 Hyclone 公司。3－(4，5－二甲基噻唑－2)－2，5－二苯基四氮唑溴盐（MTT）分析试剂盒购自中国南京基恩生物技术有限公司。

除特别声明外，所有试剂皆为分析纯且未经任何前处理。实验用水为二次去离子水（18 MΩ cm）。

2.2.3 FCDs 的制备方法

将0.2 g o－PD 加入到20 mL 甲酰胺中，超声10 分钟后，将溶液转移到50 mL 的聚四氟乙烯内胆高压釜中，在 140 ℃下加热 12 小时。冷却至室温后，将所得溶液通过0.22 μm过滤膜去除大颗粒。之后，将 100 mL 二氯甲烷加入到获得的溶液中，摇晃数下。静置15 分钟后，收集二氯甲烷相并在30 ℃旋转蒸发得到粗产物。然后，将产物溶于去离子水中，通过纤维素酯膜（MWCO：500－1 000 Da）透析 48 小时除去未反应的物质。然后收集溶液并冷冻干燥获得 CDs，将其命名为 FCDs。

为了系统研究反应溶剂对所制备 o－PD 衍生 CDs 表面态、结构和性能的影响，本研究还制备了另外三种 o－PD 衍生 CDs。具体方法如下：将 0.2 g o－PD 加入到 20 mL 水/乙醇/DMF 中，超声10 分钟后，将溶液转移到50 mL 的聚四氟乙烯内胆高压釜中，在140 ℃下加热12 小时。冷却至室温后，提纯并冷冻干燥，分别标记为 WCDs、ECDs 和 DCDs。

2.2.4 FCDs 的表征方法

通过 U－3900 紫外可见分光光度计使用1 cm 光程的比色皿，在扫描间隔为1 nm 条件下测得紫外－可见（UV－vis）吸收光谱。荧光光谱是通过 F－7000 荧光光谱仪使用1 cm光程的比色皿测得，激发和发射狭缝均设置为10 nm，扫描速度为2 400 nm · min^{-1}。原子力显微镜（AFM）图像通过使用 Bruker Dimension icon 型原子力显微镜测得，扫描模式为 Scan Asyst in air，探针型号为 Scan Asyst－air。傅立叶红外（FT－IR）光谱使用 Nicolet－6700 红外光谱仪采用溴化钾压片法测定并记录1 000～4 000 cm^{-1}的数据。X 射线光电子能谱（XPS）在配备 Al Kα 280.00 eV 激发光源的 ESCALAB 250 表面分析系统上进行测得。透射电子显微镜（TEM）图像通过 JEM－2100 高分辨透射电子显微镜测得，加速电压为200 kV。

通过参比法获得了 CDs 的量子产率。选择罗丹明6G 为参比对象（在乙醇中量子产率 =95%），发射范围为480～560 nm。首先，在 CDs 最佳激发波长下，调配不同吸光度的罗丹明6G 溶液和 CDs 溶液，它们的吸光度值不超过0.05。之后，在 CDs 最佳激发波长下，测得上述不同吸光度罗丹明6G 溶液和 CDs 溶液的荧光发射光谱，进而求得荧光发射光谱积分面积。最后，以荧光发射光谱积分面积为纵坐标，吸光度为横坐标作图，求得拟合直线斜率。采用如下公式计算量子产率：

$$\Phi_T = \Phi_R \frac{G_T}{G_R} \frac{\eta_T^2}{\eta_R^2} \qquad (2-1)$$

R 代表标准样品，T 代表测试样品，Φ 代表荧光量子产率，G 代表荧光积分面积和吸光度关系图斜率，η 代表溶剂的折光指数。

2.2.5　FCDs 的稳定性测试

pH 稳定性测试：将 FCDs 粉末分别溶于用 HCl 或 NaOH 调配后的不同 pH 值的溶液中，FCDs 在各溶液中浓度为 10 $\mu g \cdot mL^{-1}$。在 380 nm 激发波长下记录不同 pH 值样品在 545 nm 处荧光发射强度值。每组样品测试均重复 3 次。

温度稳定性测试：将 FCDs 溶液，分别置于 20 ℃、25 ℃、30 ℃、35 ℃、40 ℃、45 ℃和 50 ℃环境中孵育至少 10 分钟。在 380 nm 激发波长下记录几组样品在 545 nm 处荧光发射强度值。每组样品测试均重复 3 次。

光稳定性测试：在 365 nm 波长下连续照射 FCDs 溶液 4 000 s，并记录样品在 545 nm 处荧光发射强度值。每组样品测试均重复 3 次。

储存稳定性测试：在室温条件下将 FCDs 溶液保存 7 天，每 24 小时在 380 nm 激发波长下测试 FCDs 样品位于 545 nm 处荧光发射强度值。每组样品测试均重复 3 次。

2.2.6　FCDs 用于银离子定量检测

通过以下步骤，以 FCDs 作为荧光探针用于 Ag^+ 的定量分析。将 0.25 mL 的 Ag^+ 溶液（0 ~ 15 $\mu mol \cdot L^{-1}$）与 1.50 mL FCDs 溶液[50 $\mu g \cdot mL^{-1}$，在 pH 值为 7.0 的磷酸盐缓冲液（PBS）中]混合，所得溶液在室温下孵育 3 分钟。随后，在 380 nm 激发波长下记录溶液在 545 nm 处的荧光强度。以 Ag^+ 浓度为横坐标，$(F_0 - F)/F_0$ 为纵坐标绘制曲线，其中 F_0 和 F 分别代表加入 Ag^+ 前后 FCDs 的荧光强度。此外，在相同条件下测试了 FCDs 对其他金属离子的耐受性。所有测试均重复 3 次。

为了考察 FCDs 在实际样品中检测 Ag^+ 的可行性，本研究通过加标回收法对自然环境水样中的 Ag^+ 进行了定量分析，其中水样采集于沈阳南湖和沈阳新开河。将水样离心取上清液，用 0.22 μm 过滤膜过滤并收集滤液。以滤液为溶剂，配制了不同浓度的 Ag^+ 溶液。将 0.25 mL 不同浓度的 Ag^+ 溶液与 1.50 mL FCDs 溶液混合，所得溶液在室温下孵育 3 分钟。随后，在 380 nm 激发波长下记录溶液在 545 nm 处的荧光强度。所有测试均重复 3 次。

2.2.7 FCDs/Ag^+用于半胱氨酸定量检测

通过以下步骤,将 FCDs/Ag^+作为探针用于 Cys 的定量分析。将 0.25 mL Ag^+溶液(11 μmol·L^{-1})与 1.50 mL FCDs 溶液(50 μg·mL^{-1},在 pH 值 7.0 的 PBS 缓冲液中制备)混合,然后加入 0.25 mL 不同浓度(0~15 mmol·L^{-1})的 Cys 溶液。随后,将所得溶液在室温下孵育 3 分钟。在 380 nm 激发波长下记录 545 nm 处的荧光强度。以 Cys 浓度为横坐标,F′/F 为纵坐标绘制曲线,其中 F 和 F′分别是添加 Cys 前后 FCDs/Ag^+体系的荧光强度。此外,将 Cys 替换为干扰物,在相同条件下测试了 FCDs/Ag^+的选择性。所有测试均重复 3 次。

为了考察 FCDs/Ag^+在实际样品中检测 Cys 的可行性,本工作通过加标回收法对人血浆中的 Cys 进行了定量分析。人血浆样品从东北大学校医院获得。将乙腈加入到人血浆样品中,进行脱蛋白处理。离心并稀释后进行 Cys 检测,所有测试过程与上述方法一致。

2.2.8 FCDs 用于细胞成像

本章使用 MCF-7 细胞为研究模型,将 FCDs 用于细胞成像实验,涉及的细胞实验操作如下。

(1)细胞培养实验:

此部分主要包括 MCF-7 细胞的冻存、复苏和传代。

①冻存操作:取对数生长期的 MCF-7 细胞,倒去培养液,用 2 mL 10 mmol·L^{-1} PBS 清洗细胞 2 次,加入 1 mL 胰酶消解液,放入孵育箱中 1 分钟。将消解后不再贴壁的细胞轻轻吹洗下来,转移至离心管中,以 800 r·min^{-1}转速离心 3 分钟后,移去上清液。随后,加入 2 mL 冻存液(10% DMSO,90% FBS)将沉淀于离心管底部的细胞轻轻吹散,并将其移至冻存管中。密封后,先于 4 ℃冷却 30 分钟,再于 -20 ℃冷冻 30 分钟,最后放入超低温冰箱(-80 ℃)冻存。

②复苏操作:将保存在超低温冰箱内的 MCF-7 细胞的冻存管取出,并放置到 37 ℃水浴中解冻。待解冻后,将细胞悬浮液移至离心管内,以 800 r·min^{-1}的转速离心 3 分钟后移去上清液。将细胞接种在 4 mL 细胞培养基(添加 10% FBS、100 units·mL^{-1}青霉素和 100 mg·mL^{-1}链霉素的 DMEM 高糖培养基)中,放置于孵育箱(5% CO_2,95% 空气,湿度饱和且温度恒定为 37 ℃)中进行贴壁培养。

③传代操作:待细胞贴壁生长 36~48 h 后,进入对数生长期末期,此时可进行分瓶

操作或相应细胞实验。首先倒去培养液，用 2 mL 10 mmol · L^{-1} PBS 缓冲液清洗细胞 2 次，之后加入 1 mL 胰酶消解液，放入孵育箱中 1 分钟。将消解后不再贴壁的细胞轻轻吹洗下来，转移至离心管中，加入 1 mL 细胞培养基终止消解，以 800 r · min^{-1} 转速离心 3 分钟后移去上清液。随后，加入 1 mL 细胞培养基，将沉淀于离心管底部的细胞轻轻吹散，并将细胞悬浮液加入至细胞培养器中继续培养。

（2）细胞毒性实验：

采用标准 MTT 分析法评估了 FCDs 的体外细胞毒性。

细胞传代操作后，将 MCF－7 细胞加入到 96 孔板中间的 60 个孔中，为了保持培养环境湿润，在外围的 36 个孔中加入无菌 PBS。培养 12 小时，观察 MCF－7 细胞已基本单层铺满孔底后，移去培养基，再用 100 μL 含有不同浓度 FCDs 的 DMEM 孵育 MCF－7 细胞 24 小时。孵化结束后，移去培养基。随后，将 10 μL MTT（5 mg · mL^{-1}）加入每个孔中并进一步孵育 4 小时形成甲瓒。接下来，将 100 μL DMSO 加入每个孔以溶解活细胞产生的甲瓒。最后，使用酶标仪记录每个样品在 545 nm 处的光密度，通过公式 2－2 计算细胞存活率：

$$细胞存活率 = \frac{A_{treat}}{A_{control}} \tag{2-2}$$

A_{treat}代表加入 CDs 溶液的孔吸光度值，$A_{control}$代表空白孔吸光度值。

（3）细胞成像实验：

细胞传代操作后，将 MCF－7 细胞在含有 10% FBS 的 DMEM 培养基中培养（37 ℃，5% CO_2），待细胞进入对数生长期时，移去培养基。再用含有 100 μg · mL^{-1} 的 FCDs 培养基继续孵育 MCF－7 细胞 6 小时，随后移去含有 FCDs 的培养基，并用 PBS 缓冲液洗涤 MCF－7 细胞去除游离的 FCDs。之后，将 MCF－7 细胞与 Ag^+（25 μmol · L^{-1}）在 37 ℃下孵育 30 分钟，然后加入 Cys（25 μmol · L^{-1}）继续孵育 1 小时。与 Ag^+ 和 Cys 孵育后的 MCF－7 细胞用 405 nm 激光激发，采用共聚焦激光扫描显微镜记录成像。共焦荧光成像参数在成像分析期间保持恒定。

2.3　结果与讨论

2.3.1　FCDs 的制备与表征

反应溶剂参与的分子融合法是指：在 CDs 的制备过程中，反应溶剂和碳源之间发生了反应，参与了 CDs 的形成的一种制备方法。在本研究中，FCDs 是以 o－PD 为碳源在甲酰胺为反应溶剂的环境中通过加热碳化得到的。FCDs 的形成过程如图 2－1 所示。实际上，o－PD 上的氨基具有高活性，在高温下 o－PD 很容易通过自聚合形成聚邻苯二氨等聚合物。随后，这些聚合物上的氨基和甲酰胺上的 C＝O 发生席夫碱反应，使形成的中间产物上含有大量的 C＝N。随着反应的持续进行，中间产物持续与 o－PD 和甲酰胺反应形成最终的 FCDs。

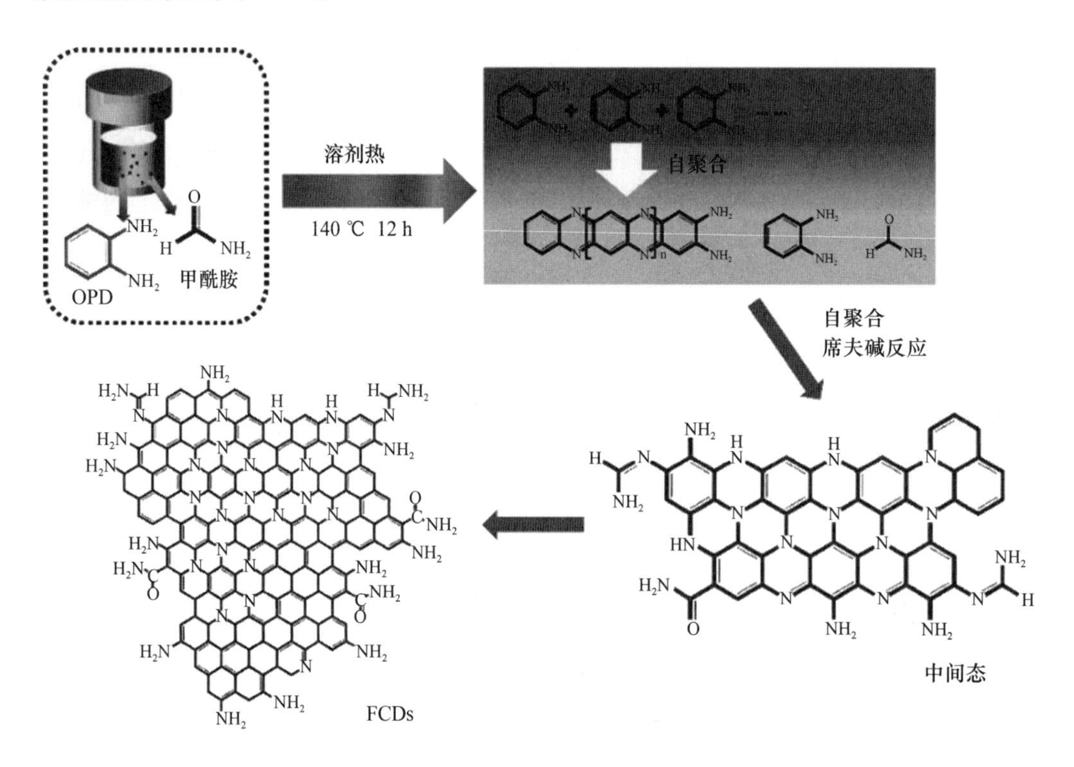

图 2－1　FCDs 形成的示意图

TEM 是把经加速和聚集的电子束投射到非常薄的样品上，电子与样品中的原子碰撞而改变方向，从而产生立体角散射。散射角的大小与样品的密度、厚度相关，因此可以形成明暗不同的影像，影像将在放大、聚焦后在成像器件（如荧光屏、胶片以及感光耦合组件）上显示出来。TEM 可以看到在光学显微镜下无法看清的小于 0.2 μm 的细微结构，这些结构称为亚显微结构或超微结构。因此，TEM 被公认为是确定 CDs 形状、形态和大小的有效手段。AFM 一种具有原子级高分辨的仪器，它可以在大气和液体环境下对各种样品进行纳米区域的物理性质、形貌进行探测，或者直接进行纳米操纵；现在 AFM 已被广泛应用于半导体、纳米功能材料、生物、化工、食品、医药研究和科研院所各种纳米相关学科的研究实验等领域中，成为纳米科学研究的基本工具。本章利用 TEM

和 AFM 对所制备的 CDs 进行了表征。图 2－2(a)中的 TEM 图像显示 FCDs 具有良好的分散性，平均尺寸为 9.7 nm，这与文献中报道的 CDs 尺寸相当[25]。高分辨率 TEM (HRTEM)图像[图 2－2(b)]显示，所制备的 FCDs 具有良好分辨的晶格条纹，其间距为 0.21 nm，这归因于石墨碳的(100)晶格距离[26]。AFM 图像[图 2－2(c)]进一步证明了 FCDs 具有良好的分散性，其高度在 7.7～8.9 nm 之间。

(a) TEM图像和尺寸分布图

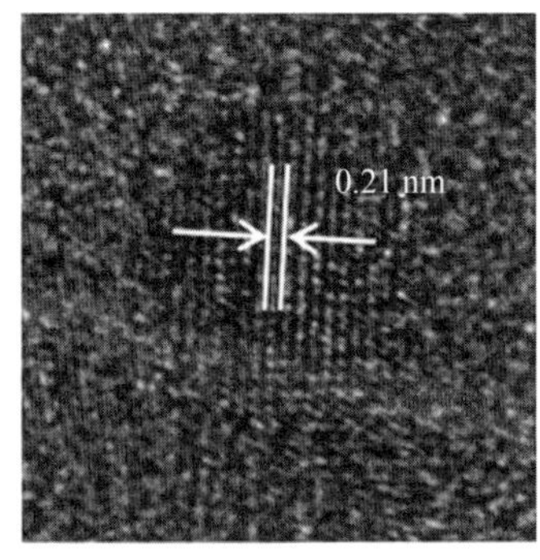

(b) HRTEM图像

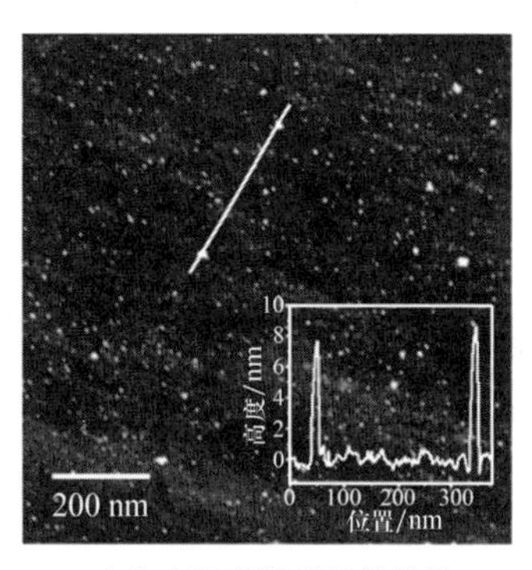

(c) AFM图像以及沿线的高度轮廓分析

图 2－2

在分子中，组成化学键或官能团的原子处于不断振动的状态，其振动频率与红外光的振动频率相当。当用红外光照射分子时，分子中的化学键或官能团可发生振动吸收，不同的化学键或官能团吸收频率不同，在 FT－IR 上将处于不同位置，从而可获得分子中含有何种化学键或官能团的信息。通过 FT－IR 光谱表征了 FCDs 上所具有的官能团。如图 2－3(a)所示，FCDs 的 FT－IR 光谱显示了如下特征吸收峰：在 3 437 cm^{-1}处为 N－H/O－H 的伸缩振动，在 2 944 cm^{-1}和 2 868 cm^{-1}处为 C－H 的伸缩振动，在 1 411 cm^{-1}处为 C＝C 的伸缩振动，在 1 340 cm^{-1}处为 C－N 的伸缩振动，在 1 259 cm^{-1}处为 N－H 的弯曲振动，以及在 1 129 cm^{-1}处 C－O 的伸缩振动[27-30]。与先前报道的以 o－PD 为前驱体制备的 CDs 相比[31]，FCDs 在 1 625 cm^{-1}处的 C＝N 伸缩振动强度更大，这个结果与甲酰胺参与的分子融合法有关。采用 XPS 进一步分析了 FCDs 的化学成分。图 2－3(b)是 FCDs 的高分辨率 XPS 光谱，从图中可以看到 C1s 谱带可以被分成 5 个结合能峰，分别对应于 C＝C/C－C(284. 2 eV)、C－N(285.2 eV)、C－O(286.0 eV)、C＝N(287. 8 eV)和－CONH－(289.0 eV)[32]。在 FCDs 的 N 1s 光谱中[图 2－3(c)]展示了 398.3 eV、399.2 eV、400.0 eV 和 401.1 eV 4 个峰，分别对应于吡啶氮、氨基氮、吡咯氮以及石墨氮[33]。

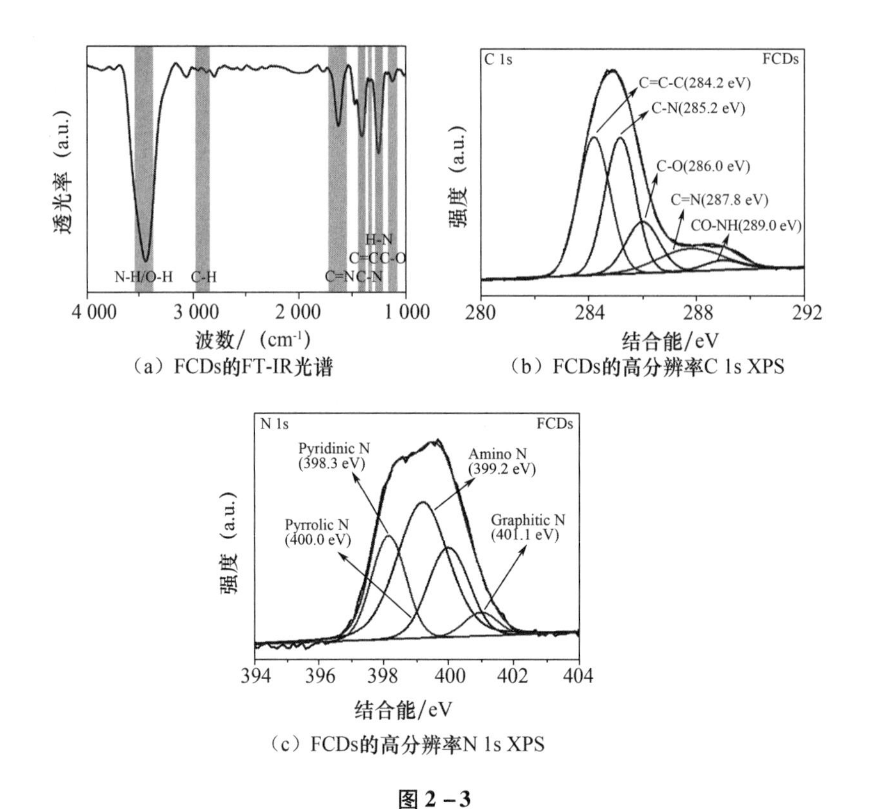

（a）FCDs的FT-IR光谱

（b）FCDs的高分辨率C 1s XPS

（c）FCDs的高分辨率N 1s XPS

图2－3

此外，本工作还表征了以其他反应溶剂制备得到的o－PD衍生CDs（WCDs、ECDs以及DCDs）的高分辨XPS光谱。如图2－4所示，与FCDs相同，另外3种CDs的高分辨C 1s XPS同样可以被分成5个结合能峰，分别对应于C＝C/C－C、C－N、C－O、C＝N和－CONH－。3种CDs的N 1s谱也同样分为4个峰，分别对应于吡啶氮、氨基氮、吡咯氮以及石墨氮。这个结果说明4种o－PD衍生CDs具有相同的官能团，即反应溶剂的改变并没有影响4种o－PD衍生CDs表面官能团的种类。

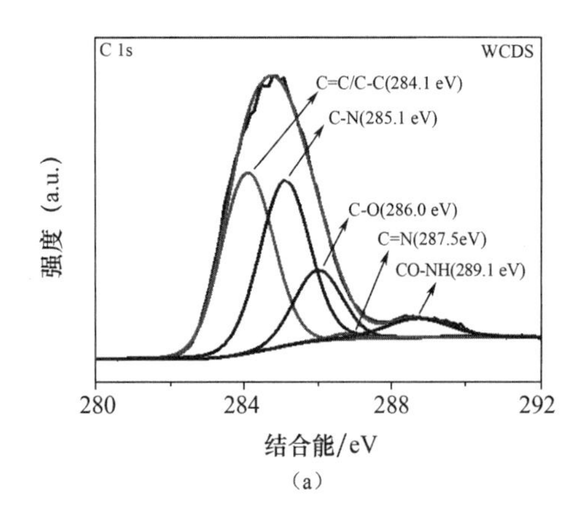

（a）

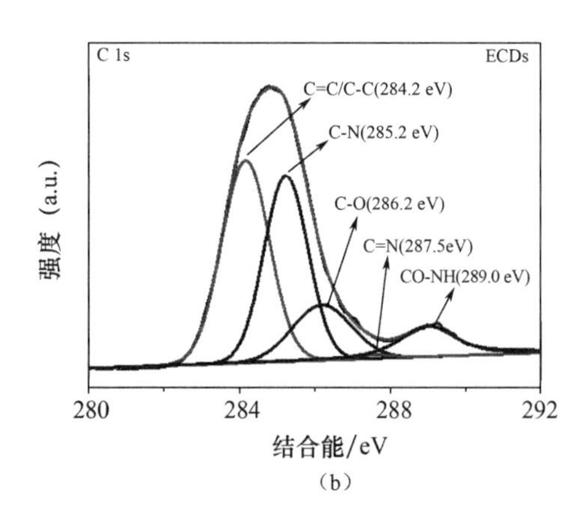

（b）

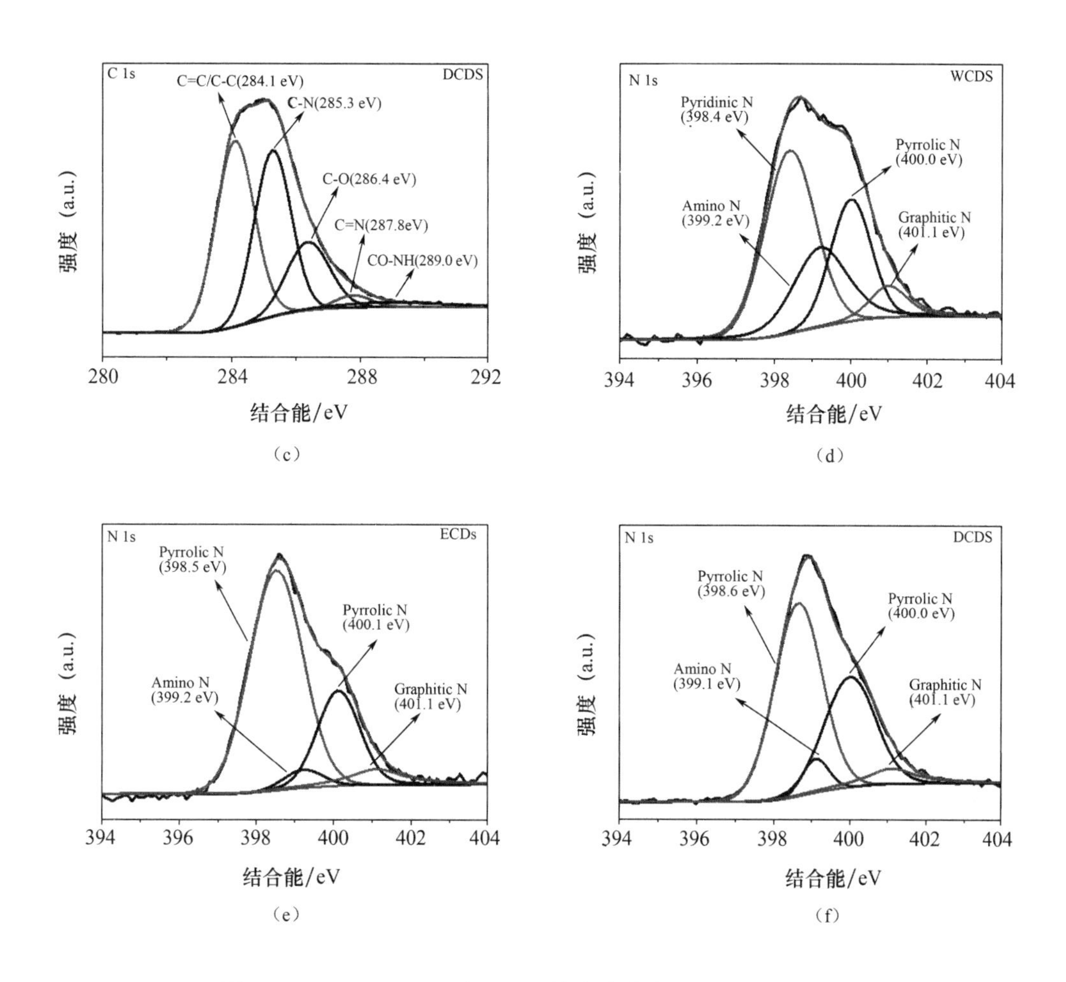

（c）　（d）　（e）　（f）

图 2－4　WCDs、ECDs 和 DCDs 的高分辨率 C 1s 和 N 1s XPS

为了获得 4 种 o－PD 衍生 CDs 中所含官能团的定量信息并确定溶剂参与的分子融合法的效果，本研究通过对 4 种 o－PD 衍生 CDs 的 XPS 分峰后相应区域进行积分，评估了采用不同溶剂制备的 o－PD 衍生 CDs（FCDs、WCDs、ECDs 以及 DCDs）表面官能团的相对含量，结果列于表 2－1 和表 2－2。从表可以看出，在 FCDs 上，由于碳源 o－PD 和反应溶剂甲酰胺之间的席夫碱反应，使 FCDs 上含有最高数量的 C＝N 基团。与之相比，由于其他三种反应溶剂并不能提供较多数量的 C＝O 基团与 o－PD 的氨基反应，导致 WCDs、ECDs 和 DCDs 上 C＝N 基团含量极低。除此之外，在 4 种 o－PD 衍生 CDs 之中，FCDs 上还具有最多数量的氨基氮，这是 FCDs 保留了碳源 o－PD 和反应溶剂甲酰胺的氨基基团所引起的。这个结果表明反应溶剂参与的分子融合法通过改变官能团数量进而影响了 4 种 o－PD 衍生 CDs 的表面态。

表 2-1　FCDs、WCDs、ECDs 和 DCDs 的 C 1s XPS 分析结果

样品名称	C=C/C-C/%	C-N/%	C-O/%	C=N/%	CO-NH/%
FCDs	36.42	34.22	14.26	12.15	2.95
WCDs	40.63	38.25	15.12	0.87	5.13
ECDs	41.34	34.86	13.90	0.66	9.24
DCDs	42.20	35.25	16.82	2.49	3.24

表 2-2　FCDs、WCDs、ECDs 和 DCDs 的 N 1s XPS 分析结果

样品名称	吡啶氮/%	氨基氮/%	吡咯氮/%	石墨氮/%
FCDs	22.51	49.06	23.45	4.98
WCDs	51.89	6.73	34.59	6.79
ECDs	65.69	3.96	23.59	6.76
DCDs	54.25	5.18	34.07	6.50

2.3.2　FCDs 的光学性质

在化合物分子中有形成单键的 σ 电子、有形成双键的 π 电子、有未成键的孤对 n 电子。当分子吸收一定能量的辐射能时，这些电子就会跃迁到较高的能级，此时电子所占的轨道称为反键轨道，而这种电子跃迁同内部的结构有密切的关系，这就产生了 UV-Vis 光谱。FCDs 的 UV-vis 吸收光谱如图 2-5 所示。从图中可以看到 FCDs 在 380 nm 处存在一个吸收带，这是 C-N/C=N 或 C-O/C=O 的 $n-\pi^*$ 跃迁所引起的[34]。与此同时，FCDs 的荧光激发和发射光谱显示（图 2-5），在 380 nm 激发下，FCDs 在 545 nm 处展现了最大发射峰。值得注意的是，FCDs 荧光激发光谱与吸收光谱在 380 nm 处的吸收峰重叠，这说明 FCDs 的荧光发射主要是由其表面含氮和氧的结构所引起的[35]。如图 2-6 所示，在 4 种 o-PD 衍生 CDs 的不同激发波长下的荧光发射光谱可以看出，4 种 o-PD 衍生 CDs 均显示出了不依赖激发波长的发射行为，这可以说明 4 种 o-PD 衍生 CDs 荧光发射均受到表面态发射调控[35,36]。另外，FCDs 的最大发射波长为 545 nm，低于其他 3 种 o-PD 衍生 CDs 的最大发射波长。这是因为石墨氮可诱导 CDs 荧光发射红移，而与其他 3 种 o-PD 衍生 CDs 相比，FCDs 中的石墨氮含量相对较低（表 2-2），所以它的发射波长也略低于其他 3 种 o-PD 衍生 CDs[37]。这种现象再一次证明了反应溶剂参与的分子融合法能够用于获得不同表面态的 CDs 制备。以罗丹明 6G 为参比，测

得 FCDs、WCDs、ECDs 和 DCDs 的量子产率分别为 20.5%、21.0%、18.3%和 18.5%。

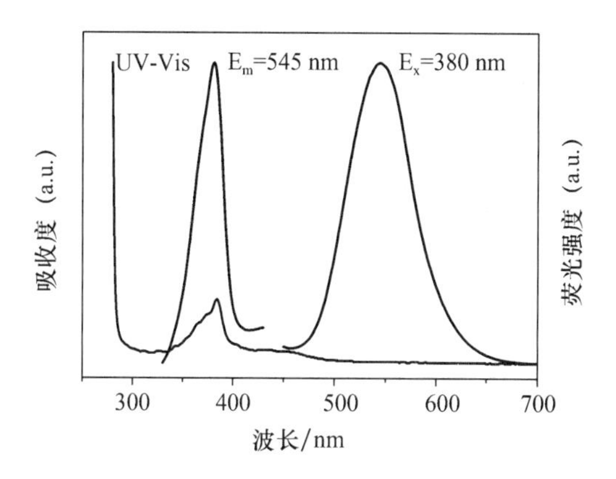

图 2-5　FCDs 的 UV-vis 吸收光谱、荧光激发光谱和荧光发射光谱

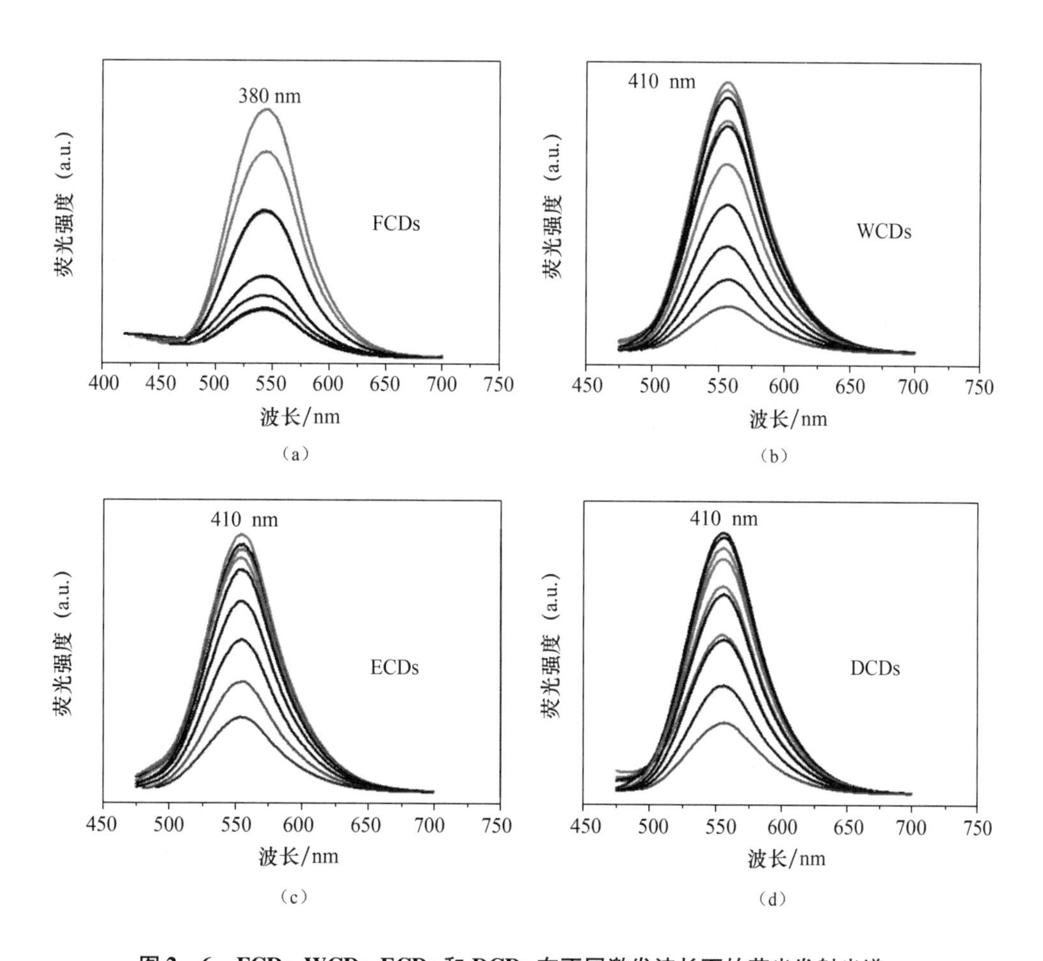

图 2-6　FCDs、WCDs、ECDs 和 DCDs 在不同激发波长下的荧光发射光谱

考察了不同 pH 环境条件下对 FCDs 的荧光发射行为的影响,结果如图 2－7(a)所示。在强酸性和强碱性环境中 FCDs 会发生荧光猝灭,但是除了强酸强碱条件外,FCDs 的荧光在较大范围内(pH 值为 4.0～8.0)能够保持相对稳定。此外,本工作还考察了不同环境温度对 FCDs 的荧光性能的影响。从图 2－7(b)可以看出,当环境温度在 20～50 ℃时,FCDs 表现出了相对稳定的荧光发射。此外,当用氙气灯连续照射 FCDs 超过 1 小时,FCDs 仍然展现出了非常稳定的荧光发射[图 2－7(c)]。而且,在室温下存放 7 天后,FCDs 依旧表现出良好的荧光稳定性[图 2－7(d)]。这些结果表明 FCDs 具有出色的稳定性,可以在周围环境条件改变的情况下展现出稳定的荧光发射。同时,这也为 FCDs 作为功能强大的荧光探针用于实际生活奠定了坚实的基础。

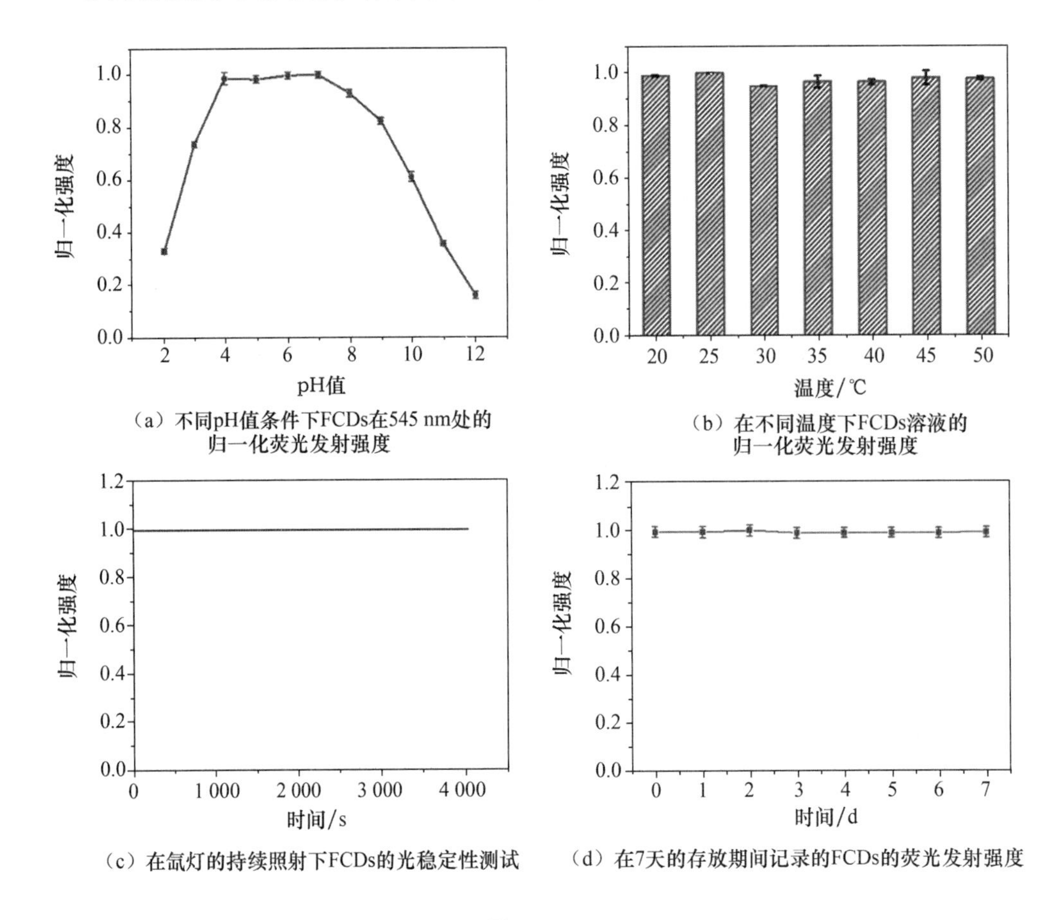

(a) 不同pH值条件下FCDs在545 nm处的归一化荧光发射强度

(b) 在不同温度下FCDs溶液的归一化荧光发射强度

(c) 在氙灯的持续照射下FCDs的光稳定性测试

(d) 在7天的存放期间记录的FCDs的荧光发射强度

图 2－7

为了考察 FCDs 在不同溶剂中的荧光发射情况。在本研究中,将 FCDs 分散在不同溶剂中并以 380 nm 为激发波长测试了 FCDs 在几种溶剂中的荧光发射光谱。如图 2－8 所示,将 FCDs 分散在四氢呋喃、丙酮、乙腈、甲醇和水后,FCDs 呈现出了较宽的颜色发射(495～545 nm),这种现象的产生归因于 CDs 和溶剂之间的非特异性(例如:偶极子极)和特异性(例如:氢键)相互作用[38]。这个发现开辟了一个简单的设计思路用于得到多色 CDs。此外,这个发现也使 CDs 能够在相应的应用中代替传统的半导体量子点,从而避免进行复杂的合成,进而降低生产成本[39]。

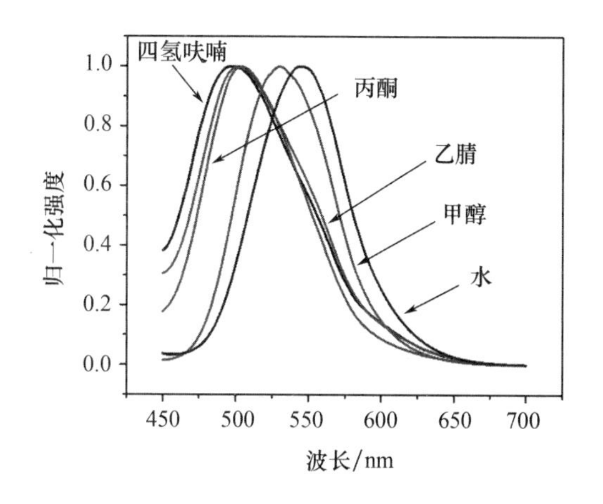

图 2－8　不同溶剂中 FCDs 的归一化荧光发射光谱

2.3.3　基于 FCDs 的银离子荧光检测

FCDs 具有的独特光学特性为其在传感领域的应用提供了可能。筛选测试揭示了 FCDs 荧光发射会因 Ag^+ 的存在而发生淬灭现象,并且所表现的荧光淬灭几乎不受 pH 改变的影响(图 2－9),这表明 FCDs 具有通过荧光“关闭”模式作为纳米探针用于 Ag^+ 检测的潜力。

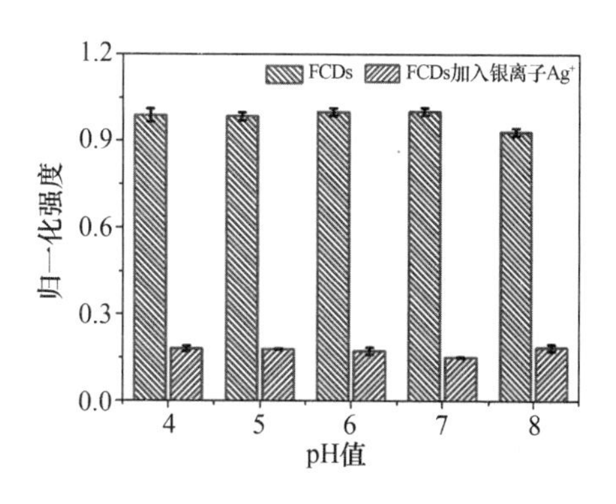

图 2－9　在不同 pH 条件下加入 Ag^+ 前后 FCDs 在 545 nm 处的归一化荧光发射强度

进一步考察了不同 Ag^+ 浓度下,FCDs 的荧光发射变化情况[图 2－10(a)]。如图 2－10(b)所示,FCDs 的荧光强度随着 Ag^+ 浓度的增加而降低,当 Ag^+ 浓度为 15 μmol·L^{-1}时,荧光淬灭率$(F_0-F)/F_0$接近 90%。当 Ag^+ 浓度在 0～15 μmol·L^{-1} 时,$(F_0-F)/F_0$ 与 Ag^+ 浓度呈正相关关系[图 2－10(b)]。在 0.05～11.00 μmol·L^{-1} 浓度范围内符合线性关系,拟合得到线性回归方程为 $(F_0-F)/F_0=0.085C-0.023$ ($R^2=0.9947$),检出限为 0.019 μmol·L^{-1}(3σ/K)。表 2－3 中总结了一些以 CDs 为探针检测 Ag^+ 的相关报道的检测性能数据,从表可以看出本工作展现出了较好的检测性能。

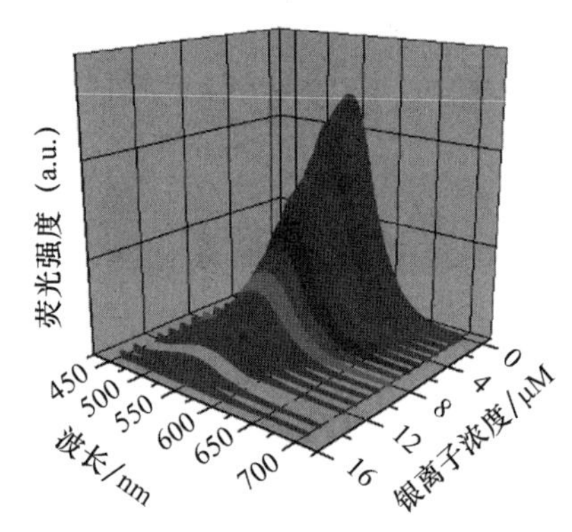

(a)加入不同浓度的Ag⁺后FCDs的荧光发射光谱

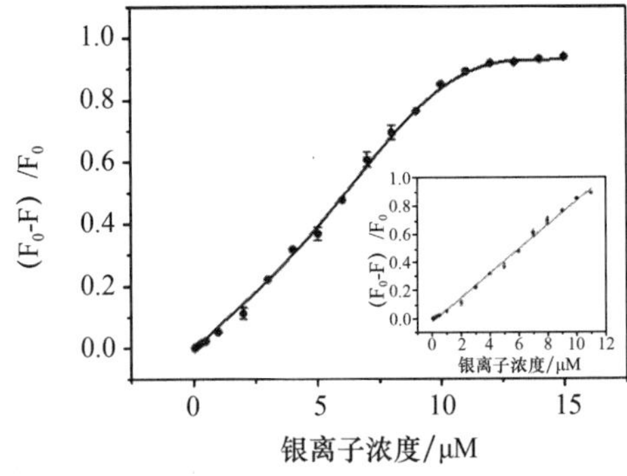

(b) $(F_0\text{-}F)/F_0$与Ag^+浓度之间的关系
[插图: $(F_0\text{-}F)/F_0$与Ag^+浓度线性相关]

图 2－10

表 2－3 使用 CDs 作为探针的 Ag^+ 检测性能比较

探针名称	线性范围/(μmol·L^{-1})	检出限/(μmol·L^{-1})	参考文献
C－dots	0－90	0.32	[40]
DNA－derived bio－dots	1－10	0.31	[41]
CQDs	0－600	0.5	[12]
S,N－CQDs	0－250	0.4	[42]
F－CDs	0－70	0.082	[43]
y－CDs	1－7	0.20	[44]
FCDs	0.05－11	0.019	本工作

为了了解 Ag^+ 对 FCDs 荧光淬灭的机理,本工作测试了 Ag^+ 加入前后 FCDs 的荧光寿命(图 2－11)。在未加入 Ag^+ 时,FCDs 的荧光寿命为 6.82 ns。加入 Ag^+ 后,FCDs 的

荧光寿命为 3.4 ns。荧光寿命缩短表示在 FCDs 和 Ag^+ 之间发生了极快的电子转移过程[45]。这可能是 CDs 上源自碳源和反应溶剂的 $-NH_2$ 基团以及碳源和反应溶剂之间通过席夫碱反应产生的 C = N 基团所引起的。一般来说，Ag^+ 与 $-NH_2$ 基团之间有螯合趋势[46]；而 Ag^+ 和 C = N 之间会发生配位作用[47]。与其他 3 种 o－PD 衍生的 CDs 相比，当向 WCDs、ECDs 和 DCDs 溶液（50 mg · mL^{-1}）中加入 10 μmol · L^{-1} Ag^+ 后，几乎没有观察到荧光发射强度的变化（图 2－12）。这是受反应溶剂的影响，在 WCDs、ECDs 和 DCDs 结构中只有很少的 $-NH_2$ 和 C = N 基团存在，这极大削弱 Ag^+ 与 CDs 之间的相互作用。这些结果表明反应溶剂参与的分子融合法在 FCDs 的制备过程中对调控表面态起到了重要作用，并影响 FCDs 的性质。

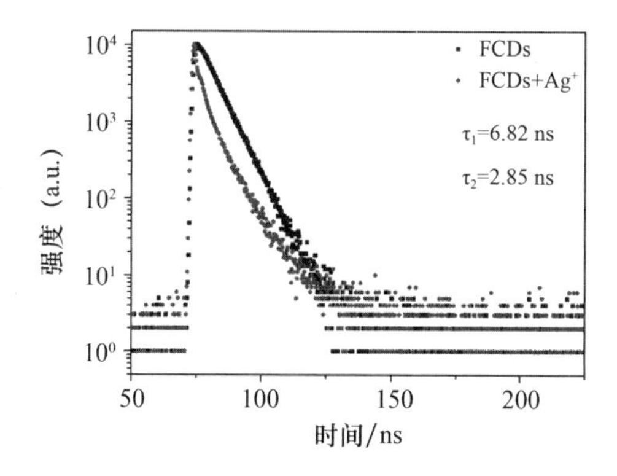

图 2－11　FCDs 和 FCDs/Ag^+ 体系的荧光寿命衰减曲线

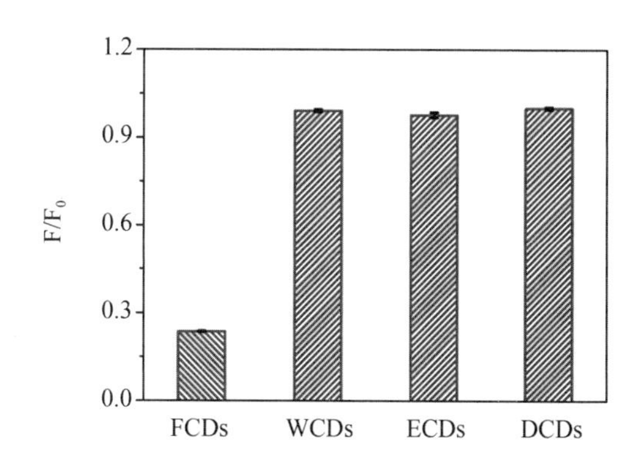

图 2－12　不同反应溶剂制备的 CDs 添加 Ag^+ 后的相对荧光强度 F/F_0

通常情况下，当金属离子达到一定浓度时，CDs 的荧光发射会受影响。因此，在本研究中，测试了 FCDs 对一些常见金属离子的耐受性。表 2－4 列出了 FCDs 的荧光光谱变化在 ±5% 的范围内对各种金属离子最大耐受浓度。从表中可以看出，FCDs 对不同的金属离子表现出了不同的耐受性。同时，FCDs 对不同金属离子的耐受性差异可以通过软硬酸碱(HSAB)理论进行解释。所谓的 HSAB 理论是指,软碱更易与软酸相互作用,而硬碱则倾向于与硬酸缔合[48]。作为 o-PD 衍生 CDs,在 FCDs 结构中含有高比例的苯环结构,使 FCDs 具有"软碱"特性[37]。Na^{+}、K^{+}、Mg^{2+}、Ba^{2+}、Li^{+}、Ca^{2+} 和 Mn^{2+} 属于硬酸,Co^{2+}、Cu^{2+}、Ni^{2+} 和 Pb^{2+} 属于临界酸。因此,FCDs 对这些离子表现出极高的耐受性。而 Au^{+}、Hg^{2+} 和 Cd^{2+} 属于软酸[49,50],因此 Au^{+}、Hg^{2+} 和 Cd^{2+} 会在一定程度上影响 FCDs 的荧光性能。值得注意的是,FCDs 对 Au^{+}、Hg^{2+} 和 Cd^{2+} 的耐受水平高于 Ag^{+} 130~1 000 倍,这表明 FCDs 对 Ag^{+} 具有良好的选择性。另外,FCDs 对常见金属离子的耐受性明显优于之前报道的 CDs(表 2-5),这说明通过反应溶剂参与的分子融合法使 FCDs 的性能得到了显著提升。

表 2-4 导致 FCDs 荧光强度发生 ±5%变化的金属离子浓度

	金属离子种类	金属离子浓度/(μmol·L⁻¹)
硬酸	Na^{+}	6.0×10^{6}
	K^{+}	4.0×10^{6}
	Mg^{2+}	2.0×10^{6}
	Ba^{2+}	1.5×10^{6}
	Li^{+}	1.5×10^{6}
	Ca^{2+}	5.0×10^{4}
	Mn^{2+}	5.0×10^{4}
临界酸	Co^{2+}	1.5×10^{3}
	Ni^{2+}	1.5×10^{3}
	Pb^{2+}	1.5×10^{3}
	Cu^{2+}	1.3×10^{3}
软酸	Cd^{2+}	1.0×10^{3}
	Hg^{2+}	3.0×10^{2}
	Au^{+}	1.3×10^{2}
	Ag^{+}	1.0

表 2-5 最近报道的 CDs 对金属离子耐受性总结

金属离子种类	FCDs 的耐受性[a]	y-CDs 的耐受性[44]	C-dots 的耐受性[51]	u-CDs 的耐受性[52]
Na^{+}	1.2×10^{5}	2.0×10^{2}	10	5
K^{+}	8.0×10^{4}	—	10	5

续表

金属离子种类	FCDs 的耐受性[a]	y－CDs 的耐受性[44]	C－dots 的耐受性[51]	u－CDs 的耐受性[52]
Mg^{2+}	4.0×10^4	—	10	5
Ba^{2+}	3.0×10^4	—	—	—
Li^{+}	3.0×10^4	—	—	5
Ca^{2+}	1.0×10^3	—	—	—
Mn^{2+}	1.0×10^3	2.0×10^{-1}	10	5
Co^{2+}	30	2.0×10^{-1}	10	5
Ni^{2+}	30	2.0×10^{-1}	—	5
Pb^{2+}	30	2.0×10^{-1}	10	5
Cu^{2+}	26	2.0×10^{-1}	10	5
Cd^{2+}	20	2.0×10^{-1}	10	5

a:金属离子与 CDs 的浓度比。

为了评价 FCDs 检测 Ag^+ 的实用性和可行性。本工作以 FCDs 为探针,检测了环境水样中 Ag^+ 的含量。如表 2－6 所示,环境水样的加标回收率范围为 99.5%～104.8%。上述结果表明 FCDs 纳米探针具有在复杂基质样品中检测 Ag^+ 的潜在能力。

表 2－6　实际样品中 Ag^+ 的测定结果

样品名称	测定量/($\mu mol\cdot L^{-1}$)	加标量/($\mu mol\cdot L^{-1}$)	回收值/($\mu mol\cdot L^{-1}$)	回收率/%
湖水	ND	0.5	0.52 ± 0.03	104.0
		2.0	1.99 ± 0.01	99.5
		6.0	6.29 ± 0.03	104.8
河水	ND	0.5	0.52 ± 0.02	104.0
		2.0	2.01 ± 0.03	100.5
		6.0	6.26 ± 0.11	104.3

2.3.4　基于 FCDs/Ag^+ 的半胱氨酸荧光检测

根据 Ag^+ 与硫醇基团之间特殊的相互作用,本研究基于荧光“开启”模式将 FCDs/Ag^+ 体系用于 Cys 检测中。如图 2－13(a)所示,FCDs/Ag^+ 体系在 545 nm 处的荧光发射强度随着 Cys 的加入逐渐上升。当 Cys 加入浓度在 0～15 $\mu mol\cdot L^{-1}$ 时,FCDs 荧光增强程度与 Cys 的浓度呈现正相关关系[图 2 - 13(b)]。当 Cys 加入浓度在 0.05 ~ 9 μmol · L^{-1}时,FCDs 荧光增强程度与 Cys 的浓度呈线性相关,线性方程为 F′/F = 0.904C + 0.975 (R^2 = 0.996 3),检出限为 0.015 μmol · L^{-1}(3σ/K)。为了进一步评估 FCDs/Ag^+体系的选择性,对添加常见生物分子(包括:氨基酸、葡萄糖、抗坏血酸和硫胺素)后 FCDs/Ag^+ 体系荧光强度变化进行了研究。如图 2 - 14 所示,当加入 500 μmol · L^{-1}干扰物后,FCDs/Ag^+体系荧光强度并未发生明显变化。应该提到的是,高浓度的其他生物硫醇(例如:GSH 和 Hcy)在传感系统中会造成一定程度的荧光恢复。但是,考虑到人血浆中 Cys 的水平远远高于 GSH 和 Hcy(20 ~ 50 倍)[53-55]。因此,FCDs/Ag^+体系可用作为纳米探针检测人血浆中的 Cys 含量。

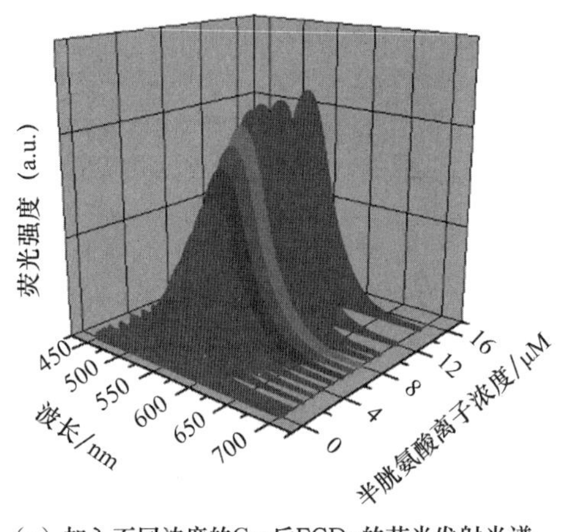

(a) 加入不同浓度的Cys后FCDs的荧光发射光谱

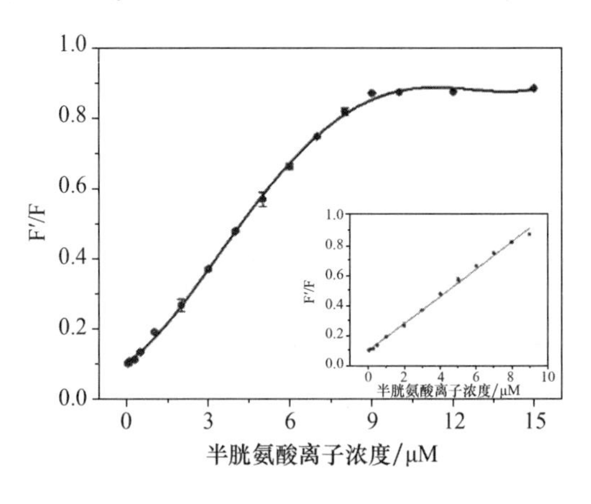

(b) F′/F与Cys浓度之间的关系
(插图:F′/F与Cys浓度线性相关)

图 2 - 13

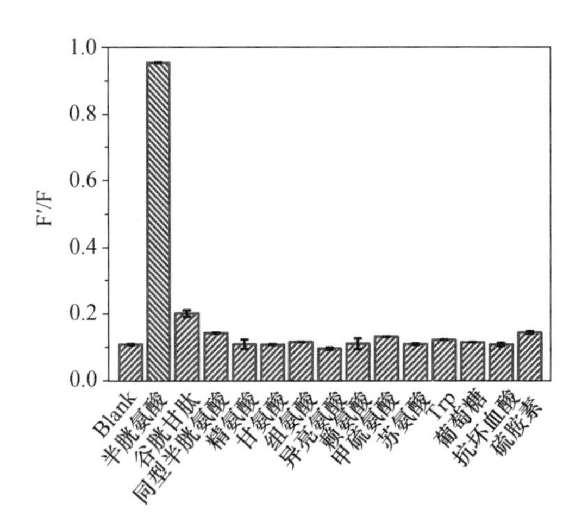

图 2 - 14　FCDs/Ag^+体系对 Cys 的选择性检测;Cys 浓度为 10 μmol · L^{-1},GSH 和 Hcy 浓度为 2 μmo · L^{-1},其他干扰物的浓度为 500 μmo · L^{-1}

为了评价 FCDs/Ag^+ 体系在实际样品中检测 Cys 可行性，本工作以 FCDs/Ag^+ 体系作为探针检测了人血浆样品中的 Cys 的含量。如表 2－7 所示，通过加标实验验证方法的准确性，可以看到人血浆的加标回收率范围为 96.5% ~104.0% 。此外，由 FCDs/Ag^+ 体系检测的人血浆中 Cys 含量与之前报道文献中 Cys 含量一致[51,53]。上述结果表明 FCDs/Ag^+ 体系具有在复杂基质样品中检测 Cys 的潜在能力。

表 2－7　实际样品中 Cys 的测定结果

样品名称	测定值/(μmol · L^{-1})	加标量/(μmol · L^{-1})	回收值/(μmol · L^{-1})	回收率/%
血浆 1	1.38 ± 0.02	2.0	3.31 ± 0.01	96.5
		4.0	5.54 ± 0.10	104.0
血浆 2	1.33 ± 0.01	2.0	3.28 ± 0.01	97.5
		4.0	5.25 ± 0.12	98.8

2.3.5　基于 FCDs 的细胞成像

本工作评估了 FCDs 纳米探针在活细胞中成像的潜力，首先通过标准 MTT 分析了 FCDs 的细胞毒性。MCF－7 细胞与 300 μg · mL^{-1}的 FCDs 孵育 24 小时后，MCF－7 细胞的存活率为 95%（图 2－15）。这证明 FCDs 具有低细胞毒性和良好的生物相容性。随后，将 FCDs 作为荧光探针用于细胞成像。首先将 MCF－7 细胞与 100 μg · mL^{-1} FCDs 孵育。然后，在共聚焦荧光显微镜下观察。从图 2－16(a) 可以清晰地看出经 FCDs 孵育后 MCF－7 黄色荧光的细胞图像。随后，外源 Ag^+（25 μmo · L^{-1}）的引入引起黄色荧光淬灭[图 2－16(b)]。用 Cys(25 μmo · L^{-1})进一步孵育 MCF－7 细胞后，黄色荧光恢复[图 2－16(c)]。这些结果表明 FCDs 可用作潜在的荧光探针用于实时监测活细胞中的 Ag^+/Cys。

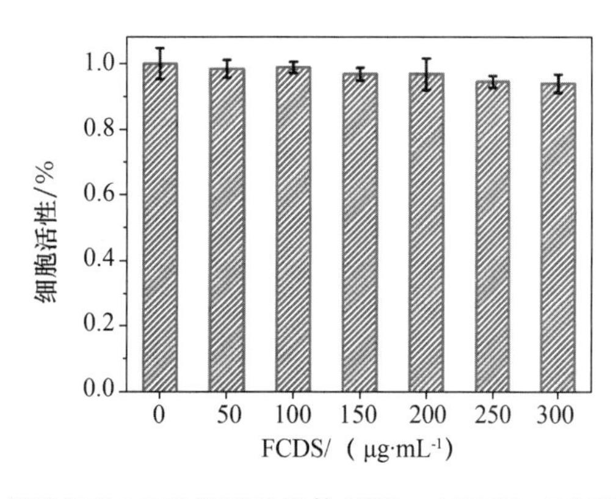

图 2－15　使用标准 MTT 测定法评估 FCDs 对 MCF－7 细胞的细胞毒性

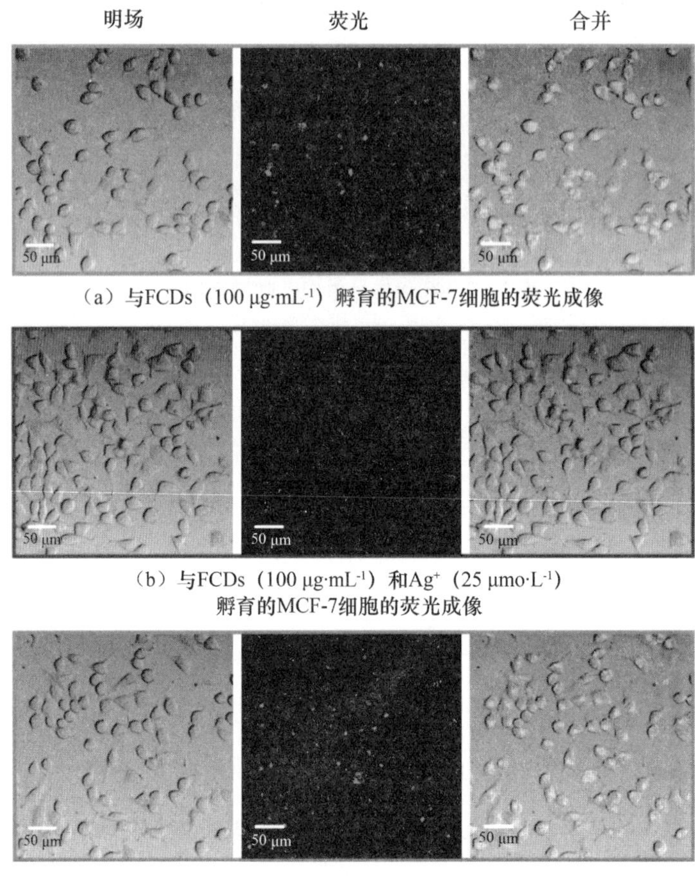

（a）与FCDs（100 μg·mL⁻¹）孵育的MCF-7细胞的荧光成像

（b）与FCDs（100 μg·mL⁻¹）和Ag^+（25 μmo·L⁻¹）孵育的MCF-7细胞的荧光成像

（c）与FCDs（100 μg·mL⁻¹），Ag^+（25 μmo·L⁻¹）和Cys（25 μmo·L⁻¹）孵育的MCF-7细胞的荧光成像

图2－16

2.4 小结

本章研究提出了一种涉及反应溶剂参与的分子融合法制备不同表面态 o－PD 衍生 CDs 的方法。其中使用甲酰胺作为反应溶剂制备得到的 FCDs 表面拥有丰富的 $-NH_2$ 和 C＝N 基团。而且，与其他 3 种 o－PD 衍生 CDs 相比，FCDs 对 Ag^+ 表现出了良好的选择性，可以用于实际样品中 Ag^+ 的定量分析。这表明反应溶剂会显著影响 o－PD 衍生 CDs 的表面态并改善其性质。同时，受本身结构的影响，FCDs 展现出了“软碱”性质，对硬酸和临界酸类金属离子表现出了极强的耐受性。此外，基于荧光“开启”模式，本工作还将 FCDs/Ag^+ 体系用于 Cys 的定量分析中，检测结果令人满意。最后，FCDs 还被开发

为细胞内 Ag^{+} 和 Cys 成像的探针。综上所述,溶剂参与的分子融合法不仅为实用性 CDs 的设计提供了一条新颖的途径。同时,也为 CDs 的表面态与性能关系优化提供了新见解。

虽然,基于反应溶剂参与的分子融合法成功用于开发不同表面态 o－PD 衍生 CDs 的制备中,并对 o－PD 衍生 CDs 的表面态/结构和性能的关系有了新的认识。但是,制备过程中涉及高温加热过程,这对环境产生了一定负担,也在一定程度上限制了 CDs 的生产。同时,目前对于表面态如何调控 o－PD 衍生 CDs 的性质还是片面的。因此,在下一章的研究工作中,希望开发一种绿色的制备不同表面态 o－PD 衍生 CDs 的方法,并进一步深入研究表面态调控对 CDs 性质影响。

参 考 文 献

[1] Sun Z, Zhou W, Luo J, et al. High－efficient and pH－sensitive orange luminescence from silicon－doped carbon dots for information encryption and bio－imaging [J]. Journal of Colloid and Interface Science, 2022, 607: 16－23.

[2] He C, Xu P, Zhang X, et al. The synthetic strategies, photoluminescence mechanisms and promising applications of carbon dots: Current state and future perspective [J]. Carbon, 2022, 186: 91－127.

[3] Guo J, Lu Y, Xie A－Q, et al. Yellow－emissive carbon dots with high solid－state photoluminescence [J]. Advanced Functional Materials, 2022: 2110393.

[4] Zhu P, Zhu Z, Li Z, et al. Nitrogen, sulfur co－doped red carbon dots for sensitive and selective detection of Sn^{2+} ions [J]. Optical Materials, 2021, 121: 111543.

[5] Zhao Y, Geng X, Shi X, et al. A fluorescence－switchable carbon dot for the reversible turn－on sensing of molecular oxygen [J]. Journal of Materials Chemistry C, 2021, 9 (12): 4300－4306.

[6] Sun S, Zhao L, Wu D, et al. Manganese－Doped Carbon Dots with Redshifted Orange Emission for Enhanced Fluorescence and Magnetic Resonance Imaging [J]. ACS Applied Bio Materials, 2021, 4 (2): 1969－1975.

[7] Liu Q, Niu X, Xie K, et al. Fluorescent Carbon Dots as Nanosensors for Monitoring and Imaging Fe^{3+} and $[HPO_4]^{2-}$ Ions [J]. ACS Applied Nano Materials, 2021, 4 (1): 190－197.

[8] Lin X, Xiong M, Zhang J, et al. Carbon dots based on natural resources: Synthesis and applications in sensors [J]. Microchemical Journal, 2021, 160: 105604.

[9] Li X, Bao Y, Dong X, et al. Dual - excitation and dual - emission carbon dots for Fe^{3+} detection, temperature sensing, and lysosome targeting [J]. Analytical Methods, 2021, 13 (37): 4246 - 4255.

[10] Zhou X, Zhao G, Tan X, et al. Nitrogen - doped carbon dots with high quantum yield for colorimetric and fluorometric detection of ferric ions and in a fluorescent ink [J]. Microchimica Acta, 2019, 186 (2): 67.

[11] Zhang Y, Yuan R, He M, et al. Multicolour nitrogen - doped carbon dots: Tunable photoluminescence and sandwich fluorescent glass - based light - emitting diodes [J]. Nanoscale, 2017, 9 (45): 17849 - 17858.

[12] Liu X, Li T, Hou Y, et al. Microwave synthesis of carbon dots with multi - response using denatured proteins as carbon source [J]. RSC Advances, 2016, 6 (14): 11711 - 11718.

[13] Lu S, Sui L, Liu J, et al. Near - infrared photoluminescent polymer - carbon nanodots with two - photon fluorescence [J]. Advanced Materials, 2017, 29 (15): 1603443.

[14] Zhang T, Zhu J, Zhai Y, et al. A novel mechanism for red emission carbon dots: Hydrogen bond dominated molecular states emission [J]. Nanoscale, 2017, 9 (35): 13042 - 13051.

[15] Zhan J, Geng B, Wu K, et al. A solvent - engineered molecule fusion strategy for rational synthesis of carbon quantum dots with multicolor bandgap fluorescence [J]. Carbon, 2018, 130: 153 - 163.

[16] Vasileiadis S, Brunetti G, Marzouk E, et al. Silver toxicity thresholds for multiple soil microbial biomarkers [J]. Environmental Science & Technology, 2018, 52 (15): 8745 - 8755.

[17] Peligro F R, Pavlovic I, Rojas R, et al. Removal of heavy metals from simulated wastewater by in situ formation of layered double hydroxides [J]. Chemical Engineering Journal, 2016, 306: 1035 - 1040.

[18] 白金娜，王亮，王明慧，等. 青霉胺稳定的铜/银双金属纳米簇制备及其在银离子检测中的应用 [J]. 发光学报，2022，43 (2): 285 - 295.

[19] Reay D S, Davidson E A, Smith K A, et al. Global agriculture and nitrous oxide emis-

sions [J]. Nature Climate Change, 2012, 2 (6): 410 – 416.

[20] Ren G, Zhang Q, Li S, et al. One pot synthesis of highly fluorescent N doped C – dots and used as fluorescent probe detection for Hg^{2+} and Ag^{+} in aqueous solution [J]. Sensors and Actuators B: Chemical, 2017, 243: 244 – 253.

[21] Wu Z, Feng M, Chen X, et al. N – dots as a photoluminescent probe for the rapid and selective detection of Hg^{2+} and Ag^{+} in aqueous solution [J]. Journal of Materials Chemistry B, 2016, 4 (12): 2086 – 2089.

[22] 张召娟，侯学振，张凌素. 半胱氨酸分析检测方法的研究进展 [J]. 广东化工，2022, 49 (16): 185 – 187.

[23] Liu T, Li N, Dong J X, et al. Fluorescence detection of mercury ions and cysteine based on magnesium and nitrogen co – doped carbon quantum dots and IMPLICATION logic gate operation [J]. Sensors and Actuators B: Chemical, 2016, 231: 147 – 153.

[24] Gao X, Du C, Zhuang Z, et al. Carbon quantum dot – based nanoprobes for metal ion detection [J]. Journal of Materials Chemistry C, 2016, 4 (29): 6927 – 6945.

[25] Liu M L, Yang L, Li R S, et al. Large – scale simultaneous synthesis of highly photoluminescent green amorphous carbon nanodots and yellow crystalline graphene quantum dots at room temperature [J]. Green Chemistry, 2017, 19 (15): 3611 – 3617.

[26] Ding H, Wei J – S, Zhang P, et al. Solvent – controlled synthesis of highly luminescent carbon dots with a wide color gamut and narrowed emission peak widths [J]. Small, 2018, 14 (22): 1800612.

[27] Shi W, Fan H, Ai S, et al. Preparation of fluorescent graphene quantum dots from humic acid for bioimaging application [J]. New Journal of Chemistry, 2015, 39 (9): 7054 – 7059.

[28] Wang N, Fan H, Sun J, et al. Fluorine – doped carbon nitride quantum dots: Ethylene glycol – assisted synthesis, fluorescent properties, and their application for bacterial imaging [J]. Carbon, 2016, 109: 141 – 148.

[29] Dai X, Zhang S, Waterhouse G I N, et al. Recyclable polyvinyl alcohol sponge containing flower – like layered double hydroxide microspheres for efficient removal of As(V) anions and anionic dyes from water [J]. Journal of Hazardous Materials, 2019, 367: 286 – 292.

[30] Wang N, Han Z, Fan H, et al. Copper nanoparticles modified graphitic carbon nitride

nanosheets as a peroxidase mimetic for glucose detection [J]. RSC Advances, 2015, 5 (111): 91302 -91307.

[31] Chen S, Liu M - X, Yu Y - L, et al. Room - temperature synthesis of fluorescent carbon - based nanoparticles and their application in multidimensional sensing [J]. Sensors and Actuators B: Chemical, 2019, 288: 749 -756.

[32] Wu Q, Wang X, Rasaki S A, et al. Yellow - emitting carbon - dots - impregnated carboxy methyl cellulose/poly - vinyl - alcohol and chitosan: Stable, freestanding, enhanced - quenching Cu^{2+} - ions sensor [J]. Journal of Materials Chemistry C, 2018, 6 (16): 4508 -4515.

[33] Ding H, Yu S - B, Wei J - S, et al. Full - color light - emitting carbon dots with a surface - state - controlled luminescence mechanism [J]. ACS Nano, 2016, 10 (1): 484 -491.

[34] Ding H, Ji Y, Wei J S, et al. Facile synthesis of red - emitting carbon dots from pulp - free lemon juice for bioimaging [J]. Journal of Materials Chemistry B, 2017, 5 (26): 5272 -5277.

[35] Sun S, Zhang L, Jiang K, et al. Toward high - efficient red emissive carbon dots: Facile preparation, unique properties, and applications as multifunctional theranostic agents [J]. Chemistry of Materials, 2016, 28 (23): 8659 -8668.

[36] Dong Y, Pang H, Yang H B, et al. Carbon - based dots co - doped with nitrogen and sulfur for high quantum yield and excitation - independent emission [J]. Angewandte Chemie International Edition, 2013, 52 (30): 7800 -7804.

[37] Hola K, Sudolska M, Kalytchuk S, et al. Graphitic nitrogen triggers red fluorescence in carbon dots [J]. ACS Nano, 2017, 11 (12): 12402 -12410.

[38] Ju B, Wang Y, Zhang Y M, et al. Photostable and low - toxic yellow - green carbon dots for highly selective detection of explosive 2,4,6 - trinitrophenol based on the dual electron transfer mechanism [J]. ACS Applied Materials & Interfaces, 2018, 10 (15): 13040 -13047.

[39] Chang E, Thekkek N, Yu W W, et al. Evaluation of quantum dot cytotoxicity based on intracellular uptake [J]. Small, 2006, 2 (12): 1412 -7.

[40] Gao X, Lu Y, Zhang R, et al. One - pot synthesis of carbon nanodots for fluorescence turn - on detection of Ag^{+} based on the Ag^{+} - induced enhancement of fluorescence

[J]. Journal of Materials Chemistry C, 2015, 3 (10): 2302 – 2309.

[41] Bian S, Shen C, Qian Y, et al. Facile synthesis of sulfur – doped graphene quantum dots as fluorescent sensing probes for Ag^{+} ions detection [J]. Sensors and Actuators B: Chemical, 2017, 242: 231 – 237.

[42] Song T, Zhu X, Zhou S, et al. DNA derived fluorescent bio – dots for sensitive detection of mercury and silver ions in aqueous solution [J]. Applied Surface Science, 2015, 347: 505 – 513.

[43] Arumugam N, Kim J. Synthesis of carbon quantum dots from Broccoli and their ability to detect silver ions [J]. Materials Letters, 2018, 219: 37 – 40.

[44] Borse V, Thakur M, Sengupta S, et al. N – doped multi – fluorescent carbon dots for ‘turn off – on’ silver – biothiol dual sensing and mammalian cell imaging application [J]. Sensors and Actuators B: Chemical, 2017, 248: 481 – 492.

[45] Miao X, Yan X, Qu D, et al. Red emissive sulfur, nitrogen codoped carbon dots and their application in ion detection and theraonostics [J]. ACS Applied Materials & Interfaces, 2017, 9: 18549 – 18556.

[46] Dang D K, Sundaram C, Ngo Y – L T, et al. One pot solid – state synthesis of highly fluorescent N and S co – doped carbon dots and its use as fluorescent probe for Ag^{+} detection in aqueous solution [J]. Sensors and Actuators B: Chemical, 2018, 255: 3284 – 3291.

[47] Bhuvanesh N, Suresh S, Prabhu J, et al. Ratiometric fluorescent chemosensor for silver ion and its bacterial cell imaging [J]. Optical Materials, 2018, 82: 123 – 129.

[48] Pearson R G. Hard and soft acids and bases, HSAB, Part I [J]. Journal of Chemical Education, 1968, 45: 581 – 587.

[49] Haji Shabani A M, Dadfarnia S, Motavaselian F, et al. Separation and preconcentration of cadmium ions using octadecyl silica membrane disks modified by methyltrioctylammonium chloride [J]. Journal of Hazardous Materials, 2009, 162 (1): 373 – 377.

[50] Minamisawa H, Okunugi R, Minamisawa M, et al. Preconcentration and determination of cadmium by GFAAS after solid – phase extraction with synthetic zeolite [J]. Analytical Sciences, 2009, 22: 709 – 713.

[51] Liao S, Zhao X, Zhu F, et al. Novel S, N – doped carbon quantum dot – based “off – on” fluorescent sensor for silver ion and cysteine [J]. Talanta, 2018, 180: 300 – 308.

[52]Zuo G, Xie A, Li J, et al. Large emission red – shift of carbon dots by fluorine doping and their applications for red cell imaging and sensitive intracellular Ag^{+} detection [J]. The Journal of Physical Chemistry C, 2017, 121 (47): 26558 –26565.

[53]Jiang K, Sun S, Zhang L, et al. Bright – yellow – emissive N – doped carbon dots: Preparation, cellular imaging, and bifunctionalsensing, [J]. ACS Applied Materials & Interfaces, 2015, 7: 23231 –23238.

[54]Wang Y, Jiang K, Zhu J, et al. A FRET – based carbon dot – MnO_2 nanosheet architecture for glutathione sensing in human whole blood samples [J]. Chemical Communications, 2015, 51 (64): 12748 –12751.

[55]Jia M – Y, Niu L – Y, Zhang Y, et al. BODIPY – based fluorometric sensor for the simultaneous determination of Cys, Hcy, and GSH in human serum [J]. ACS Applied Materials & Interfaces, 2015, 7 (10): 5907 –5914.

第3章

室温氧化融合法制备不同表面态碳点及其传感应用

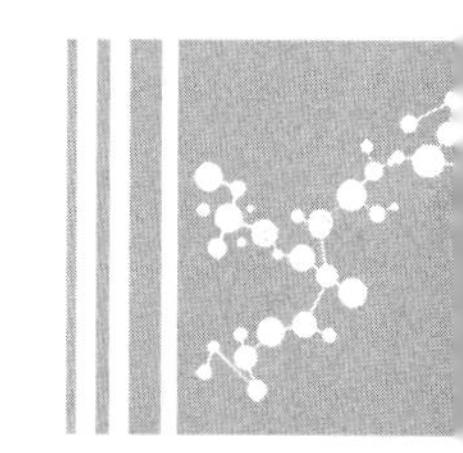

3.1 引言

作为一种强荧光纳米材料，CDs 具有低毒、优异的光学行为、优异的化学稳定性、高光稳定性和良好的生物相容性等优异特性，在过去的十年中受到了广泛的关注[1]。制备 CDs 所需的碳源来源广泛，包括：有机小分子、碳基材料、生物质甚至是食物垃圾[2-4]。所以，CDs 被视为一种满足绿色化学原理的理想纳米材料。实际上，CDs 是一种可以将应用性能最大化并最大限度地减少对环境不良影响的纳米材料[5]。在这种情况下，已经开发了许多制备 CDs 的方法，包括：水热法、溶剂热法、微波法、激光辐照法以及电化学氧化法等[6-8]。然而，这些方法都需要消耗大量的能量来达到制备 CDs 的效果。在此情况下，室温下制备 CDs 因为其制备条件简单且能耗低的优势逐渐进入人们的视野。

随着绿色化学理念的普及，新型绿色制备 CDs 工艺亟待被开发。在此背景下，室温 CDs 的研究旨在以低能耗为切入视角，为 CDs 绿色制备工艺的开发提供新的思路。此外，室温 CDs 还表现出诸多不同于传统 CDs 的特殊性质，这也引起了人们的关注。因此，研究室温 CDs 存在重要的学术和实际意义。对于室温 CDs 的制备方法，化学氧化是最常见的一种方式[9,10]。Peng 等人利用浓硫酸作为氧化剂对碳源进行脱水处理，随后又通过硝酸处理使其分解成独立的碳纳米颗粒[11]。最后，利用端氨基化合物将碳纳米颗粒钝化制备得到 CDs。Xiao 等人报道了一种以聚乙二醇作为碳源在室温下制备 CDs 的方法[12]。通过改变 NaOH 的用量，分别得到了蓝色、黄色、橘红色和红色荧光发射的 CDs。此外，他们还将此 CDs 作为荧光探针用于 Fe^{2+} 定量分析。然而，上述的制备过程大多数是需要过量的氧化剂参与的。但在后期提纯中，过量的氧化剂是难以彻底去除的，这也导致此类制备方法对环境产生了极大的危害[10,13]。因此，迫切需要开发出一种真正的在室温下简单绿色制备 CDs 的方法。

对硝基苯酚（p－NP）是一种重要的硝基芳族化合物，常用于染料、炸药、农药和药物的制造中[14,15]。然而，p－NP 是最危险和毒性最大的酚类化合物之一，它可通过急性吸入或食入引起发绀、嗜睡、头痛和恶心[15]。考虑到其高毒性，p－NP 已经被列为极度危害有毒污染物。美国环境保护局规定饮用水中 p－NP 的最大允许浓度为 0.43 $\mu mol \cdot L^{-1}$[15]。因此，非常需要开发一些用于选择性检测和监测实际样品中 p－NP 含量的方法。但是，当前报道的方法通常需要复杂的样品预处理、烦琐的电极修饰和耗时的检测程序，这限制了它们的广泛应用[16-18]。因此，建立一种经济有效、简便快捷的方法来准确测定实际样品中的痕量 p－NP 具有重要意义。

重水(D_2O)是利用其高沸点由天然水反复蒸馏产生的。在核工业、光谱表征和化学分析中起着重要作用[19,20]。但是,由于 D_2O 的高吸湿性,它很容易被 H_2O 污染,这会影响 D_2O 的进一步使用[21]。因此,许多分析方法已用于定性和定量检测 D_2O 中 H_2O 的含量,例如:原子吸收光谱法、核磁共振光谱法和红外激光光谱法等[19]。2016 年,Bohidar 等人根据铜铟镓硒纳米晶体与 D_2O 或 H_2O 之间的结合能不同,将其用作 D_2O/H_2O 传感器[22]。2017 年,Humphrey 等人利用 O－D 或 O－H 振荡器对镧系金属有机骨架中激发态镧离子具有不同的淬灭效率,将其用作 D_2O/H_2O 传感器[23]。2019 年,Zheng 和 Lu 等人基于更实用的设计策略(即筛选 pKa 大于 H_2O 的 pH 值,且小于 D_2O 的 pD 值的荧光团),开发了多种 D_2O/H_2O 分子探针[19]。当前,D_2O/H_2O 光学传感器的例子仍十分有限,要全面理解信号产生机制具有相当大的挑战性。其中,荧光传感器作为光学传感器的一种,因其操作简单、灵敏度高、即时方便以及检出限低等特点而受到越来越多的关注,但由于 H_2O 的严重干扰,用于检测 D_2O 的荧光传感器报道仍然很少。因此,有必要开发一种对 D_2O 和 H_2O 具有不同响应进而用于选择性检测 D_2O 中 H_2O 含量的荧光传感器。

在本章的研究中,采用室温氧化融合法,以 o－PD 和对苯二酚(HQ)为碳源同时绿色制备了两种 o－PD 衍生 CDs。通过硅胶色谱柱提纯,得到了两种具有不同荧光发射的 CDs(YCDs 和 GCDs)。研究发现,这两种 CDs 产物具有不同的表面态。GCDs 表面具有丰富的 $-NH_2$ 基团,而 YCDs 则富含 －OH 和 $-NO_2$ 基团。高含量的 －OH 基团使 YCDs 具有强极性。同时,丰富的 －OH 和 $-NO_2$ 基团引发 YCDs 带隙减小以及荧光发射红移。不同的表面态为两种 o－PD 衍生 CDs 提供了不同的光谱特性,并影响了它们在传感中的应用。其中,YCDs 的荧光激发光谱与污染物 p－NP 的紫外吸收范围重叠,因此内滤效应使 YCDs 成为 p－NP 荧光传感中的灵敏探针,线性范围为 0.2 ~ 50 $\mu mol \cdot L^{-1}$,检出限低至 0.08 $\mu mol \cdot L^{-1}$。而富含 $-NH_2$ 基团的 GCDs 的荧光会受到 O－H 和 O－D 不同作用而发生变化,因此可以作为荧光传感器用于分析 D_2O 中的 H_2O 含量,线性范围为 0.5% ~40% ,检出限为体积分数 0.17。

3.2 实验部分

3.2.1 实验仪器

本章所使用的仪器品牌和型号如下:

KQ－100B/800KDE 型超声波清洗器(中国昆山市超声仪器有限公司)；

BSA22AS 单盘型分析电子天平(中国北京赛多利斯仪器有限公司)；

TG16－WS 台式高速离心机(中国湘仪实验室仪器开发有限公司)；

2XZ－2 型真空泵(中国临海市谭式真空设备有限公司)；

PB－10 标准型 pH 计(中国北京赛多利斯仪器有限公司)；

Lambda Bio20 紫外可见光分光光度计(美国珀金埃尔默仪器有限公司)；

F－7000 荧光分光光度计(日本日立公司)；

JEM－2100 透射电子显微镜(日本电子株式会社)；

Bruker Dimension icon 原子力显微镜(德国布鲁克公司)；

DZF－6020 型真空干燥箱(中国上海精宏实验设备有限公司)；

One Spectra 红外光谱仪(美国珀金埃尔默仪器有限公司)；

DHG－9037A 电热恒温干燥箱(中国上海精宏实验设备有限公司)；

EscaLab 250Xi X 射线光电子能谱分析仪(美国赛默飞世尔公司)；

YE5A44 型手动可调式移液器(中国上海大龙医疗设备有限公司)；

CHI660E 电化学工作站(中国上海晨华仪器有限公司)；

EscaLab 250Xi X 射线光电子能谱分析仪(美国赛默飞世尔公司)；

Malvern Nano－ZS 粒度仪(英国马尔文公司)；

D8 ADVANCE X 射线衍射仪(德国布鲁克公司)；

Horiba XploRA 光谱仪(法国 Jobin Yvon 公司)。

3.2.2　实验试剂

本章所用化学试剂和品牌如下：

O－PD、HQ、D_2O、p－NP、邻硝基苯酚(o－NP)以及本章使用的其他芳香族化合物购自阿拉丁化学有限公司(中国上海)。磷酸、硼酸、乙酸、氢氧化钠、盐酸、硝酸银($AgNO_3$)以及本章使用的其他无机盐购自国药集团化学试剂有限公司(中国上海)。

除特别声明外,所有试剂皆为分析纯且未经任何前处理。实验用水为二次去离子水(18 MΩ cm)。

3.2.3　YCDs 和 GCDs 的制备方法

将 HQ(0.1 g)和 o－PD(0.3 g)加入到 30 mL 水中,超声直至形成透明的均匀溶液。在室温下反应 18 小时后,获得棕色悬浮液,并通过冷冻干燥得到粗产物的固体粉末。

然后使用二氯甲烷和甲醇的混合物(20:1)作为洗脱剂将粗产物用硅胶色谱柱法纯化。通过旋转蒸发除去洗脱剂,并在真空干燥箱中进一步干燥。最后,两种纯化的 CDs 产率为质量分数 21 和 15。根据获得的 CDs 的不同荧光发射,将两种 CDs 分别标记为 YCDs 和 GCDs。

3.2.4 YCDs 和 GCDs 的表征方法

荧光光谱是通过 F-7000 荧光光谱仪使用 1 cm 光程的比色皿测得,激发和发射狭缝均设置为 10 nm,扫描速度为 2 400 nm · min^{-1},三维荧光光谱扫描速度为 12 000 nm · min^{-1}。X 射线衍射(XRD)通过 D8 ADVANCE X 射线衍射仪测得,扫描范围为 10° ~50° (2θ)。拉曼(Raman)光谱是在 Horiba XploRA 光谱仪上测得。Zeta 电位使用 Malvern Nano-ZS 粒度仪在中性条件下测得。

通过循环伏安法测试了两种 CDs 的电化学性质。其中以玻璃碳电极作为工作电极,铂丝作为辅助电极,Ag/AgCl 作为参比电极。在 0.1 mol · L^{-1} 四正丁基六氟磷酸铵的乙腈溶液中测试了 CDs 样品的循环伏安曲线。CDs 的 HOMO 和 LUMO 能级根据以下公式计算得出:

$$E_{LUMO} = -e(E_{red} + 4.4) \tag{3-1}$$

$$E_{HOMO} = E_{LUMO} - E_9 \tag{3-2}$$

E_{LUMO} 代表最低未占据分子轨道的能级,E_{HOMO} 代表最高占据分子轨道的能级,E_{red} 代表还原电位的起始值,E_g 代表带隙。

此外,本章通过 U-3900 紫外可见分光光度计使用 1 cm 光程的比色皿,在扫描间隔为 1 nm 条件下测得紫外-可见(UV-vis)吸收光谱。原子力显微镜(AFM)图像通过使用 Bruker Dimension icon 型原子力显微镜测得,扫描模式为 Scan Asyst in air,探针型号为 Scan Asyst-air。傅里叶红外(FT-IR)光谱使用 Nicolet-6700 红外光谱仪采用溴化钾压片法测定并记录 1 000 ~4 000 cm^{-1} 的数据。X 射线光电子能谱(XPS)在配备 Al Kα 280.00 eV 激发光源的 ESCALAB 250 表面分析系统上进行测得。透射电子显微镜(TEM)图像通过 JEM-2100 高分辨透射电子显微镜测得,加速电压为 200 kV。

3.2.5 YCDs 和 GCDs 的稳定性测试

光稳定性测试:在 365 nm 波长下连续照射 YCDs 和 GCDs 溶液 3 600 s,并记录两种 CDs 样品在最大荧光发射时的强度值。每组样品测试均重复 3 次。

储存稳定性测试:在室温条件将 YCDs 和 GCDs 溶液下保存 7 天,并每 24 小时在最佳激发波长下测试两种 CDs 样品在最大荧光发射时的强度值。每组样品测试均重复 3 次。

3.2.6　YCDs 用于对硝基苯酚定量检测

将 YCDs 作为荧光探针用于 p－NP 的定量分析，操作过程如下：将不同浓度的 p－NP 溶液与 1.5 mL 的 YCDs 溶液混合（10 μg · mL^{-1}，在 pH 值为 9.0 的 BR 缓冲溶液中制备）并定容至 2 mL。测试 410 nm 的激发波长下各组样品的荧光光谱。随后，以 p－NP 浓度为横坐标，荧光淬灭率为 $(F_0-F)/F_0$ 为纵坐标绘图，其中 F_0 和 F 分别代表添加 p－NP 前后 YCDs 的荧光强度。此外，在相同条件下将 p－NP 替换为干扰物研究了 YCDs 的选择性。所有测试均重复 3 次。

为了考察 YCDs 在实际样品中检测 p－NP 的可行性，本研究以 YCDs 作为探针通过加标回收法对环境水样中的 p－NP 进行了定量分析。环境水样采集于沈阳南湖和沈阳新开河。将水样离心取上清液，用 0.22 μm 过滤膜过滤并收集滤液。以滤液为溶剂，配制不同浓度 p－NP 溶液。将不同浓度 p－NP 溶液与 1.5 mL YCDs 溶液混合并定容至 2 mL。随后，在 410 nm 激发下测试样品荧光发射光谱。所有测试均重复了 3 次。

3.2.7　GCDs 用于重水中水定量检测

以 GCDs 作为荧光探针用于 D_2O 中 H_2O 含量检测，操作过程如下：将 25 mg 的 GCDs 粉末分散在 20 mL 的 D_2O 中，然后超声处理 15 分钟以确保溶液的均匀性。然后，将不同体积的 H_2O 添加到 50 μL GCDs 溶液中。随后，用 D_2O 定容至 2 mL。在 440 nm 的激发波长下测试样品的荧光发射光谱。以 D_2O 中的 H_2O 含量为横坐标，荧光强度比 F/F_0 为纵坐标作图，其中 F_0 和 F 分别代表 GCDs 的 D_2O 溶液中和 GCDs 在具有不同 H_2O 含量 D_2O 中的荧光强度。此外，在相同条件下向反应体系中加入干扰物研究了 GCDs 选择性。所有测试均重复 3 次。

为了考察 GCDs 用于实际样品中检测 D_2O 中 H_2O 含量的可行性，本研究以 GCDs 作为探针通过加标回收法对市售 D_2O 产品中 H_2O 含量进行了定量分析。

3.3　结果与讨论

3.3.1　YCDs 和 GCDs 的制备与表征

目前，人们已经开发出多种制备室温 CDs 的方法。例如，超声辅助酸/碱催化法：

2011 年，Kang 等人以葡萄糖为原料，在氢氧化钠（NaOH）或者盐酸（HCl）溶液中分散后，接着在室温下超声处理 4 小时制得 CDs[24]。光催化/光驱动法：2012 年，Sun 等人报道了以 o－PD 为原料，室温下在水溶液中分散后通过紫外光照射 30 分钟制备 CDs[25]。2018 年，Yan 等人以三噻吩衍生两亲化合物为原料，在水溶液中自组装后室温下通过可见光照射 16 小时制得 CDs[26]。在本工作中，研究者采用了室温氧化融合法，室温氧化融合法是指：在室温下通过前驱体之间的自发的氧化反应制备 CDs 的一种方法。在本研究中，两种 o－PD 衍生 CDs 是通过以 HQ 和 o－PD 作为前驱体，在无须任何特殊设备及外加氧化剂条件下制备得到的。CDs 的形成路线如图 3－1 所示。HQ 在空气中容易被氧化并转化为对苯醌[27]。同时，o－PD 的氨基具有高活性，因此在空气中容易通过氧化/聚合形成各种氧化产物，包括 o－PD 二聚体、o－PD 三聚体和 2，3－二硝基吩嗪甲烷等[28,29]。在此，席夫碱反应发生在 o－PD 氧化产物的 $-NH_2$ 与对苯醌的 C＝O 之间。此外，席夫碱反应还释放出少量的热量促进 o－PD 和 HQ 的聚合和氧化反应[27]。随着反应的进行，中间产物继续与 o－PD 氧化产物和对苯醌反应形成 CDs。由于 HQ 上的－OH 基团和 o－PD 上的 $-NH_2$ 基团都容易被氧化，因此两个前驱体的氧化反应处于竞争状态，导致 CDs 形成过程中出现两种不同的途径，并最终得到具有不同表面态的 CDs。在路线 I 中，o－PD 氧化/聚合主导 CDs 的形成，所以更多 o－PD 上 $-NH_2$ 基团被氧化为 $-NO_2$ 基团，而 HQ 上的－OH 基团被保留至 CDs 上，使制备的 YCDs 具有丰富的 $-NO_2$ 和－OH 基团。而在途径 II 中，HQ 的氧化在 CDs 的形成中起主要作用，因此 HQ 上的－OH 基团被消耗，较多 o－PD 的 $-NH_2$ 基团被保留在了 GCD 上。

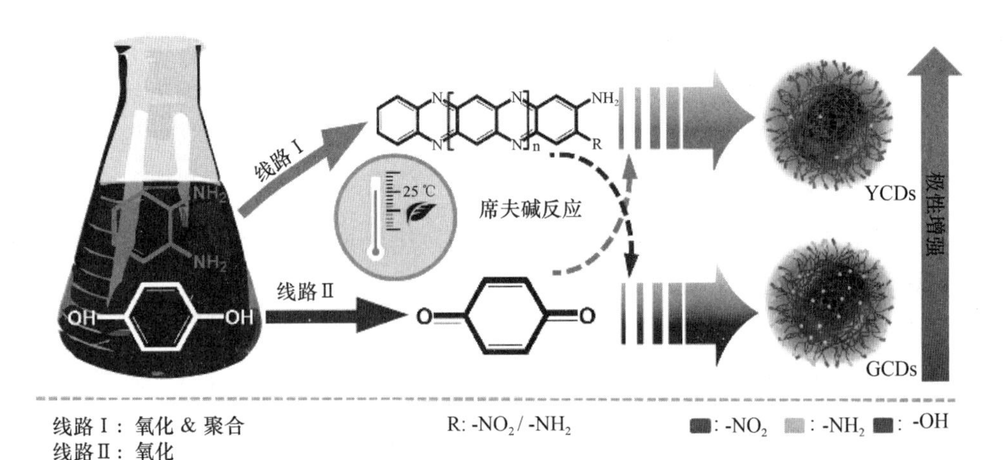

图 3－1　CDs 形成示意图

在许多研究中，CDs 是通过透析纯化得到的[30]。但是，这些纯化的 CDs 的荧光发射光谱很宽，并且显示出了激发波长依赖的荧光发射行为[31]。通常，透析只能通过尺寸效应筛选出 CDs，并且所得产物仍是具有不同物理化学性质的复杂混合物 CDs[32]。硅胶色谱柱法是最近报道的一种 CDs 纯化技术，它摒弃了依据尺寸的大小来纯化 CDs 的思路，而是根据 CDs 物理化学性质差异进行提纯。因此，通常可获得具有不依赖激发波长的荧光行为和不同荧光发射的 CDs 样品[33]。本研究中，使用硅胶色谱柱法通过优化洗脱液组成和配比，发现所获得的产物是具有不同荧光发射的 CDs 的混合物。如图 3 - 2所示，粗产物可以很好地分为两种产物，这表明这两种 CDs 的极性差异很大。

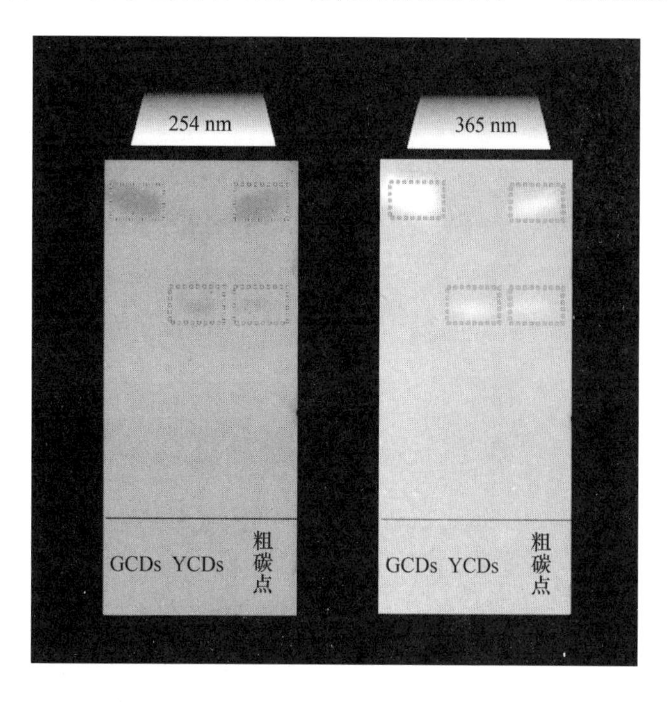

图 3 - 2　YCDs、GCDs 和粗 CDs 的薄层色谱图

TEM 是把经加速和聚集的电子束投射到非常薄的样品上，电子与样品中的原子碰撞而改变方向，从而产生立体角散射。散射角的大小与样品的密度、厚度相关，因此可以形成明暗不同的影像，影像将在放大、聚焦后在成像器件（如荧光屏、胶片、以及感光耦合组件）上显示出来。TEM 可以看到在光学显微镜下无法看清的小于 0.2 μm 的细微结构，这些结构称为亚显微结构或超微结构。因此，TEM 被公认为是确定 CDs 形状、形态和大小的有效手段。AFM 是一种具有原子级高分辨的仪器，它可以在大气和液体环境下对各种样品进行纳米区域的物理性质、形貌进行探测，或者直接进行纳米操纵；现在 AFM 已被广泛应用于半导体、纳米功能材料、生物、化工、食品、医药研究和科研院所各种纳米相关学科的研究实验等领域中,成为纳米科学研究的基本工具。在本章中,TEM 和 AFM 测试用于表征所制备 CDs 的形状、大小和结构。图 3-3(a)和 3-3(b)的 TEM 图像显示两种类型的 CDs 均是单分散的,且平均粒径分别为 20.7 nm 和 20.9 nm。它们都是圆球形的纳米点,分布均匀且无团聚。[图 3-3(c)和 3-3(d)]相应的 AFM 图像表明平均高度约为 20 nm。由 AFM 确定的 CDs 的平均大小与 TEM 结果非常吻合。

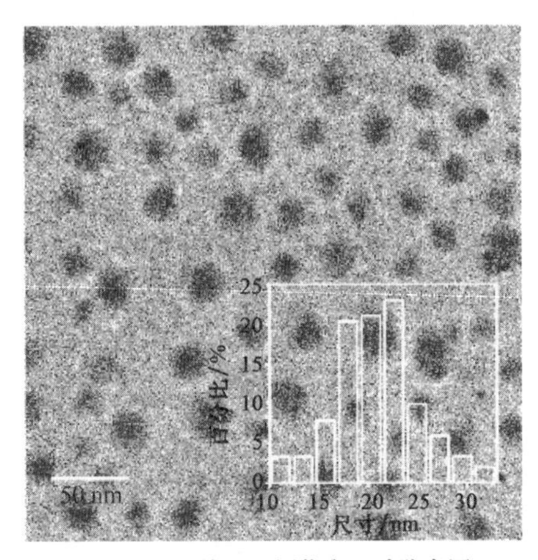

(a) GCDs的TEM图像和尺寸分布图

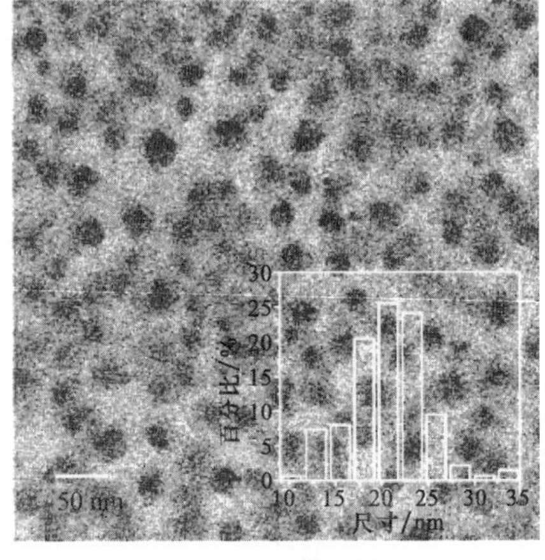

(b) YCDs的TEM图像和尺寸分布图

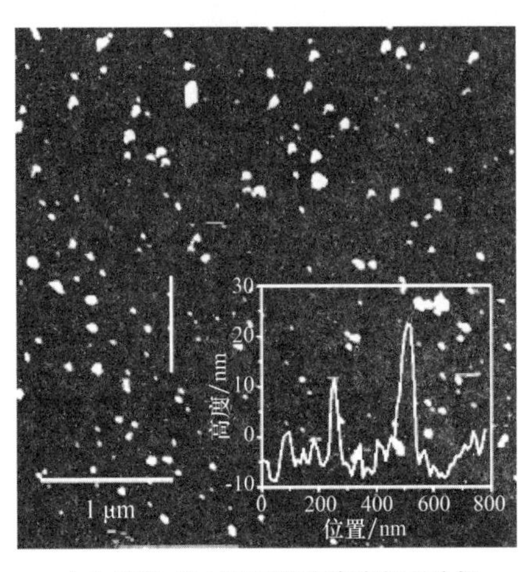

(c) GCDs的AFM图像和高度剖面分析

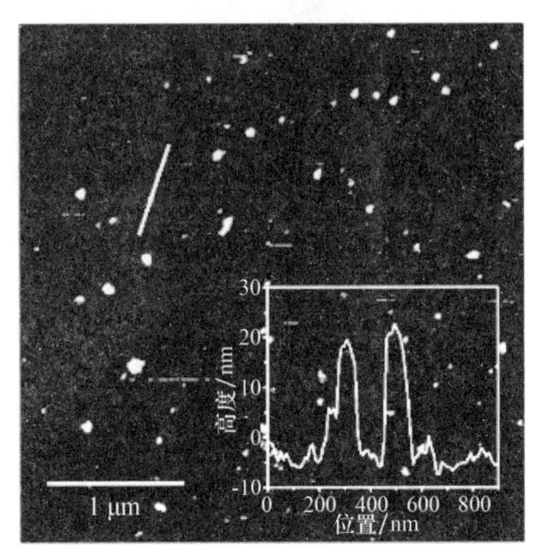

(d) YCDs的AFM图像和高度剖面分析

图 3-3

在分子上组成化学键或官能团的原子处于不断振动的状态,其振动频率与红外光的振动频率相当。当用红外光照射分子时,分子中的化学键或官能团可发生振动吸收,不同的化学键或官能团吸收频率不同,在 FT-IR 上将处于不同位置,从而可获得分子中含有何种化学键或官能团的信息。利用 FT-IR 光谱对两种 CDs 上的官能团进行了表征。图 3-4(a)展示了 YCDs 的 FT-IR 光谱图,从图中可以观察到 YCDs 上存在的一些特征吸收带，在 3 400 ~ 3 500 cm^{-1}处为 O – H 的伸缩振动，在 3 200 ~ 3 400 cm^{-1}处为 N – H 的伸缩振动，在 2 932 cm^{-1}和 2 852 cm^{-1}为 C – H 的伸缩振动，在 1 625 ~ 1 635 cm^{-1}处为 C = N 的伸缩振动，在 1 500 ~ 1 510 cm^{-1}处为 C = C 的伸缩振动，在 1 405 cm^{-1}处是 C – N 的伸缩振动，在 1 330 ~ 1 360 cm^{-1}处是 – NO_2 的伸缩振动，在 1 227 cm^{-1}处为 N – H 的平面弯曲振动，以及在 1 125 ~ 1 130 cm^{-1}处 C – O 的伸缩振动[34,35]。在 GCDs 的 FT – IR 谱图也观察到了相同的吸收带，例如：O – H（3 400 ~ 3 500 cm^{-1}）、N – H（3 200 ~ 3 400 cm^{-1}）、C – H（2 923 cm^{-1}和 2 852 cm^{-1}）、C = N（1 625 ~ 1 635 cm^{-1}）、C = C（1 500 ~ 1 510 cm^{-1}）、C – N（1 405 cm^{-1}）、– NO_2（1 330 ~ 1 360 cm^{-1}）、N – H（1 227 cm^{-1}）和 C – O（1 125 ~ 1 130 cm^{-1}）。FT – IR 光谱表明，两种 CDs 是由包含氧和氮的芳族结构组成。此外，FT – IR 光谱证实了 YCDs 和 GCDs 中存在席夫碱结构（C = N 谱带），进一步揭示了两种 CDs 的形成有席夫碱反应参与。同时，在 YCDs 和 GCDs 的 FT – IR 光谱中，在 1 330 ~ 1 360 cm^{-1}处出现 – NO_2 吸收峰，这来自于 o – PD 上氨基的氧化[28]。对于两种 CDs 的氮相关基团，GCDs 的 N – H 谱带的强度要比 YCDs 的强，这表明 GCDs 从 o – PD 保留的氨基比 YCDs 多。此外，在 YCDs 的 FT – IR 光谱中 – NO_2 和 C = N 基团的峰更明显，表明 YCDs 中存在更多的 – NO_2 和 C = N 基团。此外，观察到从 GCDs 到 YCDs 的 O – H 谱带增强，这表明 YCDs 的极性比 GCDs 更高，因为宽的 O – H 谱带表明 CDs 的表面形成了分子内氢键[36]。极性差异与硅胶柱上这两种 CDs 的洗脱顺序一致。通过 XPS 光谱分析 YCDs 和 GCDs 的组成和表面态，在 YCDs 和 GCDs 的 XPS 中，位于 284 eV、399 eV 和 532 eV 处 3 个峰，分别归属于 C 1s、N 1s 和 O 1s［图 3 – 4（b）］[37 – 39]。这些结果表明样品由相同的元素组成。

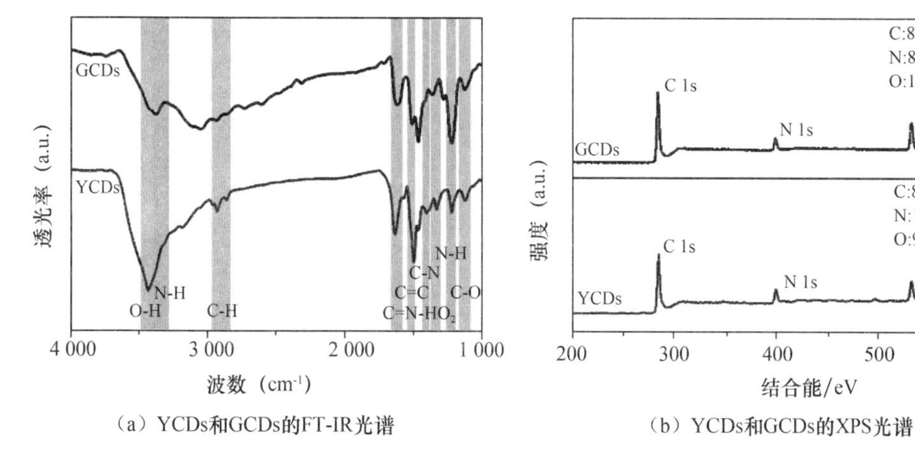

（a）YCDs和GCDs的FT-IR光谱　（b）YCDs和GCDs的XPS光谱

图 3 – 4

利用高分辨率 XPS 进一步研究两种 o－PD 衍生 CDs 的化学组成。YCDs 的 C 1s 谱带[图 3－5(a)]可以分成 4 个峰,分别对应 C＝C/C－C(284.4 eV)、C－N(285.1 eV)、C－OH(286.0 eV)和 C＝N/C＝O(288.5 eV)[40－42]。GCDs 的 C 1s 谱带[图 3－5(b)]同样可以为 4 个结合能峰,分别对应于 C＝C/C－C(284.5 eV)、C－N(285.1 eV)、C－OH(286.2 eV)和 C＝N/C＝O(288.5 eV)。在 YCDs 的 N 1s 光谱中[图 3－5(c)],出现了 3 个峰分别对应于 C－N＝C(398.9 eV)、$-NH_2$(400.0 eV)和 $-NO_2$(406.5 eV)[33,43,44]。如图 3－5(d)所示,GCDs 的 N 1s 光谱在 398.7、399.9 和 406.4 eV 处显示 3 个峰,分别对应于 C－N＝C、$-NH_2$ 和 $-NO_2$。在此,$-NO_2$ 基团的存在与这两个 CDs 的 FT－IR 光谱的结果一致。YCDs 的高分辨率 O 1s 光谱[图 3－5(e)]可以分为 531.9 eV 和 532.8 eV处的两个峰,分别对应于 C＝O 和 C－O[44]。GCDs 的 O 1s 谱带分别在 531.9 和 532.8 eV 处存在两个峰,对应 C＝O 和 C－O。

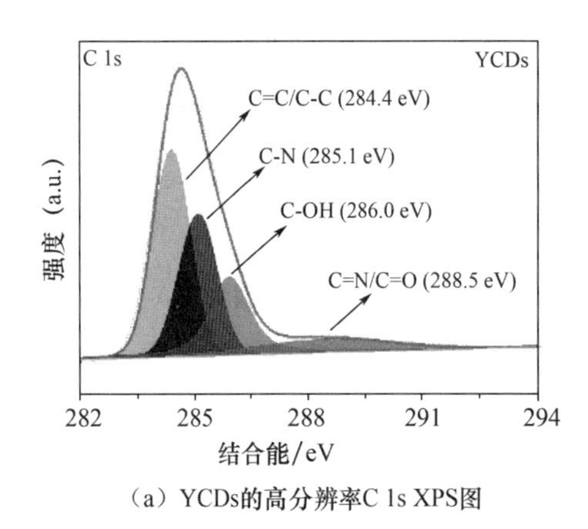

(a) YCDs的高分辨率C 1s XPS图

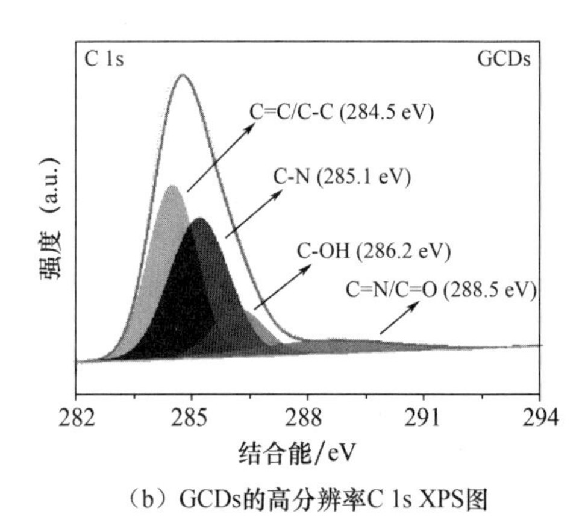

(b) GCDs的高分辨率C 1s XPS图

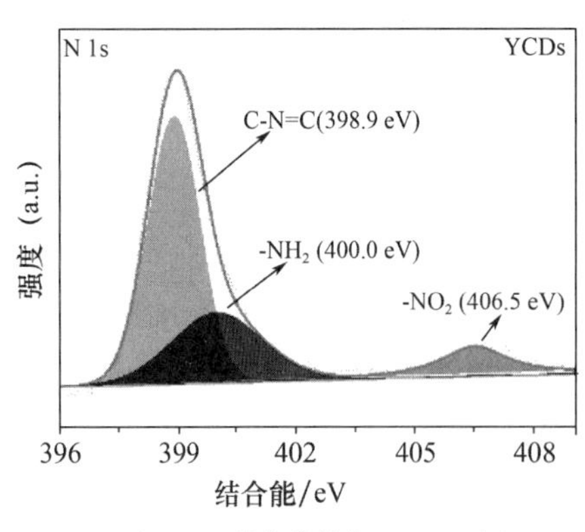

(c) YCDs的高分辨率N 1s XPS图

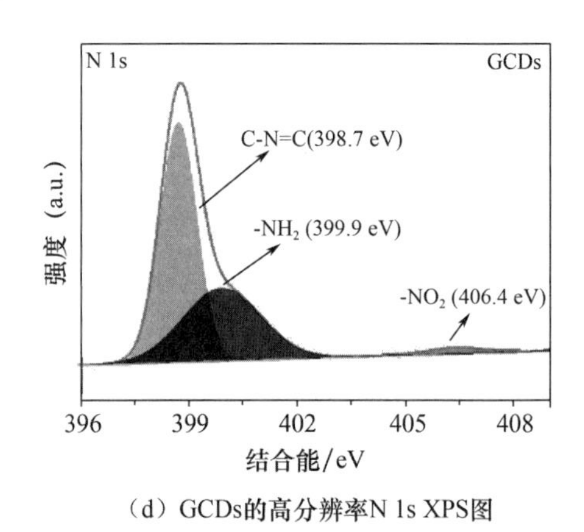

(d) GCDs的高分辨率N 1s XPS图

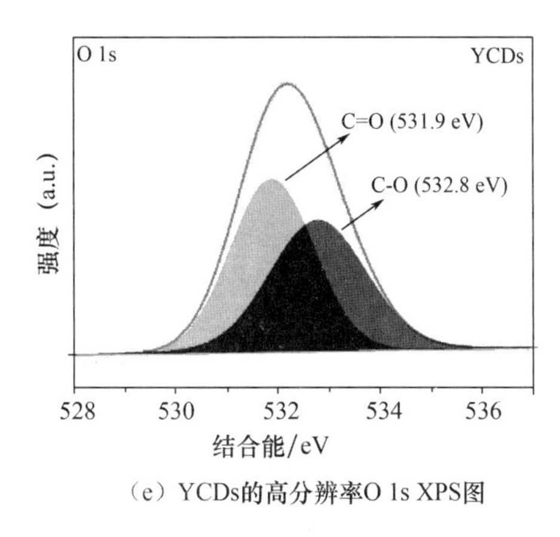

（e）YCDs的高分辨率O 1s XPS图

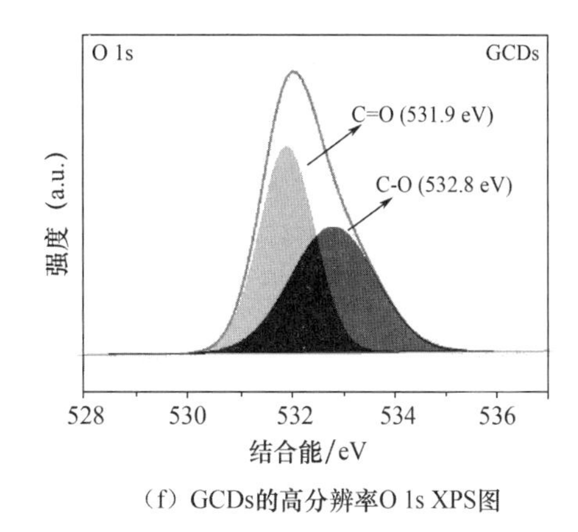

（f）GCDs的高分辨率O 1s XPS图

图 3 – 5

表 3 – 1 列出了两种 CDs 表面官能团的相对含量，这些化学官能团相对含量由 XPS 中相应的峰面积积分计算得出。从表中可以看出 YCDs 的表面上存在更多的 – OH 基团，这将赋予 YCDs 强极性[45]，这与 FT – IR 光谱的表征结果一致。因此，可通过硅胶色谱柱从粗 CDs 产品中分离出 YCDs。表面电荷分析表明，YCDs 在中性条件下显示电位为 –7.23 mV，而相同条件下测试 GCDs 的 Zeta 电位显示其略带正电荷（ +0.42 mV）。CDs 的表面电荷不同与氨基和羟基的存在相关[46]，这与 FTIR 和 XPS 光谱中表征结果一致。YCDs 和 GCDs 的 XRD 光谱都在 $2\theta = 22°$ 处显示出宽的衍射峰[图 3 – 6(a)]，这对应于 CDs 中的 sp^2 杂化碳，这是碳纳米材料的典型特征峰[47]。拉曼光谱中的无序结构或缺陷（D 带）和石墨碳域（G 带）被称为碳材料的分子指纹。两种 o – PD 衍生 CDs 的拉曼光谱[图 3 – 6(b)]在 1 374 cm^{-1} 和 1 530 cm^{-1} 处显示两个峰，分别对应于 D 带和 G 带[48]。D 带的高强度表明在这两种 CDs 样品的 sp^3 缺陷中存在局部 sp^2 簇。YCDs 的 I_G/I_D 值明显高于 GCDs，这表明 YCDs 比 GCDs 具有更高程度的石墨化程度和少量缺陷，这也是造成其量子产率高于 GCDs 的原因[44]。

表 3 – 1　两种 CDs 的 C 1s 和 N 1s 光谱 XPS 数据分析

	化学键	在 YCDs 中含量/%	在 GCDs 中含量/%
C 1s	C = C/C – C	40.65	40.30
	C – N	29.27	39.55
	C – OH	21.14	11.94
	C = N/C = O	8.94	8.21

续表

	化学键	在 YCDs 中含量/%	在 GCDs 中含量/%
N 1s	C－N＝C	59.53	58.36
	$-NH_2$	27.42	37.39
	$-NO_2$	13.05	4.25

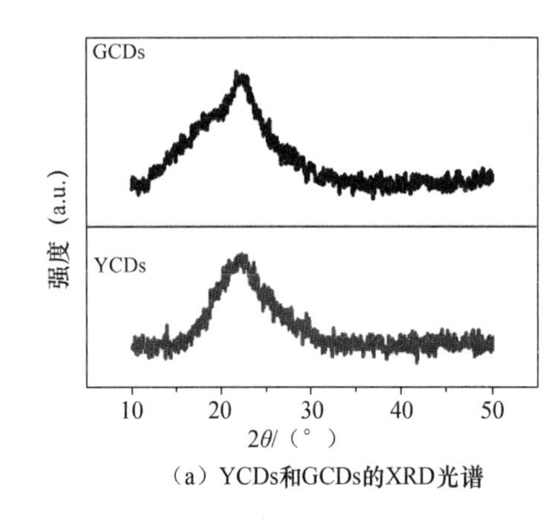

（a）YCDs和GCDs的XRD光谱

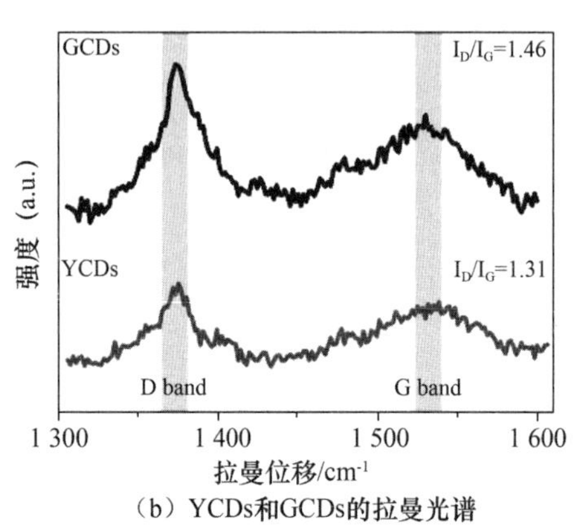

（b）YCDs和GCDs的拉曼光谱

图 3－6

上述表征说明这两种 o－PD 衍生 CDs 的官能团彼此不同，导致它们的表面态存在差异。官能团分布的差异可能归因于以下原因：前驱体中－OH 和 $-NH_2$ 基团都容易被氧化，因此在 CDs 制备过程中这两个基团之间存在竞争关系。当前驱体 o－PD 充当反应中心时，$-NH_2$ 基团的氧化反应更加活跃，因此在 YCDs 上产生了大量的 $-NO_2$ 基团。同理，当 HQ 成为反应中心时，－OH 基团的氧化反应更活跃，并且可以从 GCDs 上保留更多 $-NH_2$ 基团，这使 GCDs 中的 $-NH_2$ 基团含量高。通过元素含量可以进一步确定 CDs 不同的形成途径。从 XPS 结果得出的 YCDs 和 GCDs 的氮碳比分别为 0.126 和 0.1。YCDs 的较高 N 含量表明更多的 o－PD 参与 YCDs 的形成。相反，更多的 HQ 参与了 GCDs 的形成从而降低了氮碳比。

3.3.2 YCDs 和 GCDs 的光学性质

碳点的荧光性能无论对其在基础研究和实际应用中都是一个重要的性能。图 3－7 展示了两种 CDs 的荧光发射表征结果。图 3－7(a) 和 3－7(b) 是这两种 CDs 的 3D 荧光图，从图中可以看出 YCDs 和 GCDs 都拥有一个发射中心，YCDs 和 GCDs 的发射最大值分别在 555 nm 和 546 nm。以罗丹明 6G 为参考，YCDs 和 GCDs 的量子产率分别为

22.4% 和 16.9%。此外,[图 3-7(c)和图 3-7(d)]在不同激发波长下,两种 CDs 均表现出不依赖激发波长的荧光发射行为,这种荧光行为是由表面态发射引起的[40]。

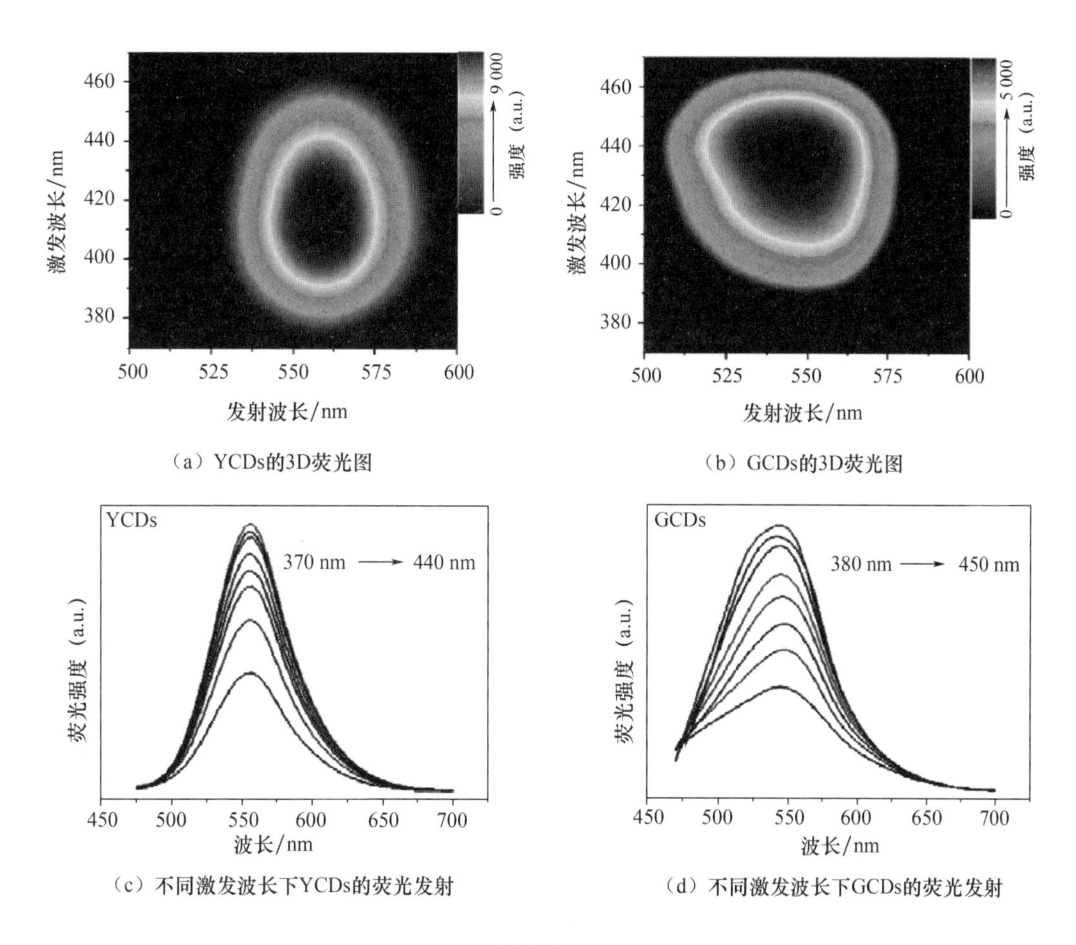

(a) YCDs的3D荧光图

(b) GCDs的3D荧光图

(c) 不同激发波长下YCDs的荧光发射

(d) 不同激发波长下GCDs的荧光发射

图 3-7

在分子中有形成单键的 σ 电子、有形成双键的 π 电子、有未成键的孤对 n 电子。当分子吸收一定能量的辐射能时,这些电子就会跃迁到较高的能级,此时电子所占的轨道称为反键轨道,而这种电子跃迁同内部的结构有密切的关系,这就产生了 UV-Vis 光谱。图 3-8 是两种 CDs 的 UV-vis 光谱、荧光激发光谱以及荧光发射光谱。从 UV-vis 光谱可以看出,在紫外光区域两种 CDs 都存在明显的吸收峰,YCDs 在 258 nm 处显示出吸收峰,而 GCDs 在 253 nm 处观察到 1 个峰,在 285 nm 处观察到肩峰,这些峰是由芳环结构中 C=C 键的 $\pi-\pi^*$ 跃迁引起的,而且这些跃迁通常不产生荧光[33]。此外,YCDs 在 420 nm 处显示特征吸收峰,GCDs 在 365 nm 和 415 nm 处显示两个吸收峰,两种 CDs 在较低能量区域所显示出不同的吸收带说明它们具有不同的表面态[37]。在 YCDs 和 GCDs 的荧光激发和发射光谱中,在 555 nm 发射下,YCDs 在 410 nm 处表现出最大激发峰。当发射波长为 546 nm 时,GCDs 的最大激发波长为 440 nm。另外,发现两个样品的相应荧光激发曲线位置接近于较低能量区域中的吸收带。这说明两种 CDs 样品的荧光发射源自低能量区域的吸收[33]。两种 CDs 的荧光激发光谱不同,表明它们的不同发射峰源自不同的发射状态[41]。为了更直观的感受两种 CDs 荧光发射的差异,在本研究中计算了两种 CDs 的色温和色坐标。在图 3 –9 中,色温为 4 552 K 和 5 713 K 的 YCDs 和 GCDs 的 CIE 色坐标分别为(0.396 9,0.589 3)和(0.321 3,0.590 3)。

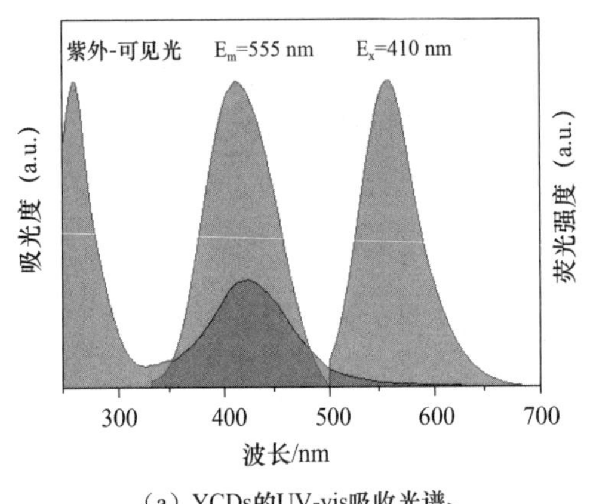

(a) YCDs的UV-vis吸收光谱、荧光激发光谱和荧光发射光谱

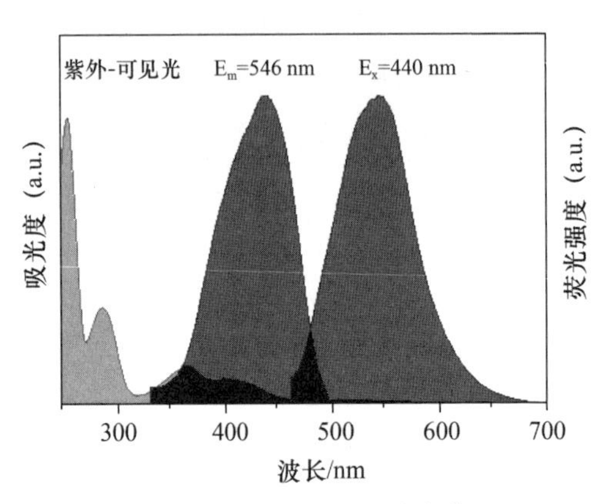

(b) GCDs的UV-vis吸收光谱、荧光激发光谱和荧光发射光谱

图 3 –8

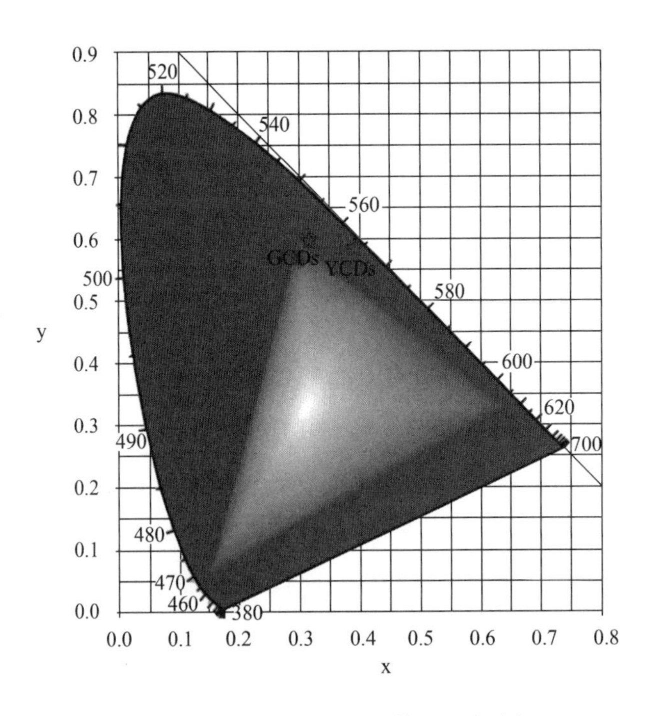

图 3 –9 YCDs 和 GCDs 的 CIE 色坐标

为了更深入地了解这两种 CDs 的荧光特性，本研究测试了 CDs 的荧光寿命[图 3－10(a)和图 3－10(b)]。两种 CDs 的荧光衰减曲线可以通过单指数函数拟合，YCDs 和 GCDs 的寿命分别为 2.48 ns 和 1.75 ns。通常，单指数荧光寿命表明样品包含单重态荧光中心[32]。此外，不同的荧光寿命进一步证实了不同类型的电子激发态和两种 CDs 的不同性质[49]。

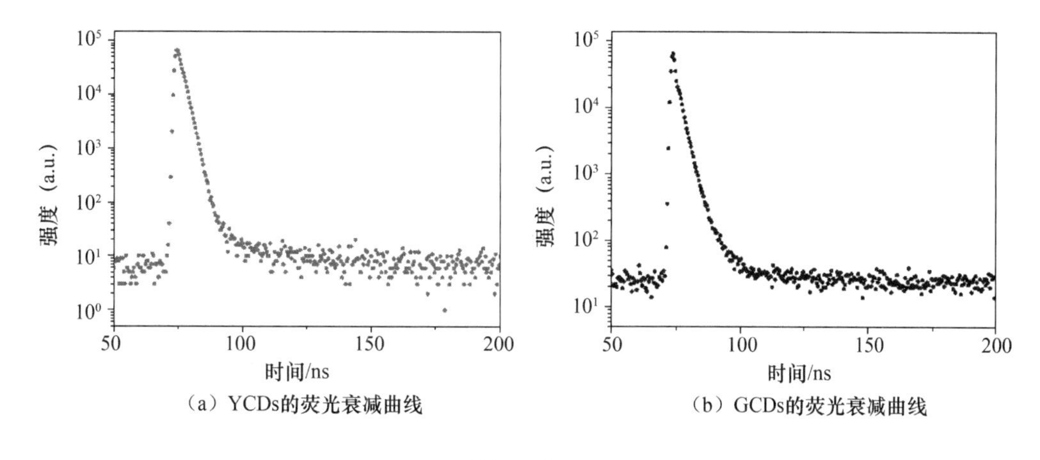

(a) YCDs的荧光衰减曲线　　(b) GCDs的荧光衰减曲线

图 3－10

在不同 pH 值条件下测试 CDs 的荧光发射(图 3－11)，结果显示不同 pH 值环境中 YCDs 和 GCDs 的荧光变化存在极大差异，这种 pH 值响应的差异源自于 YCDs 和 GCDs 之间的表面态差异[40]。

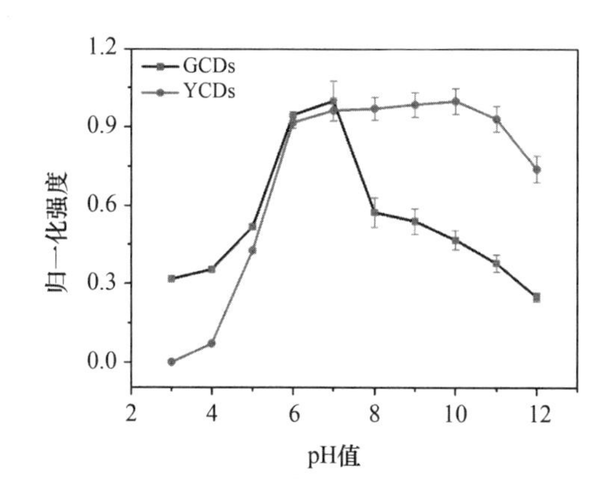

图 3－11　不同 pH 值下的 YCDs 和 GCDs 的归一化荧光强度

对于两种 CDs 的光稳性测试显示，在氙灯连续照射 3 600 s 下，YCDs 和 GCDs 的荧光强度几乎没有变化，这说明两种 CDs 均具有极高的光稳定性[图 3－12(a)和

图3－12(b)]。此外，YCDs和GCDs在室温下保存7天后仍展现出了良好的稳定性[图3.12(c)和图3－12(d)]。

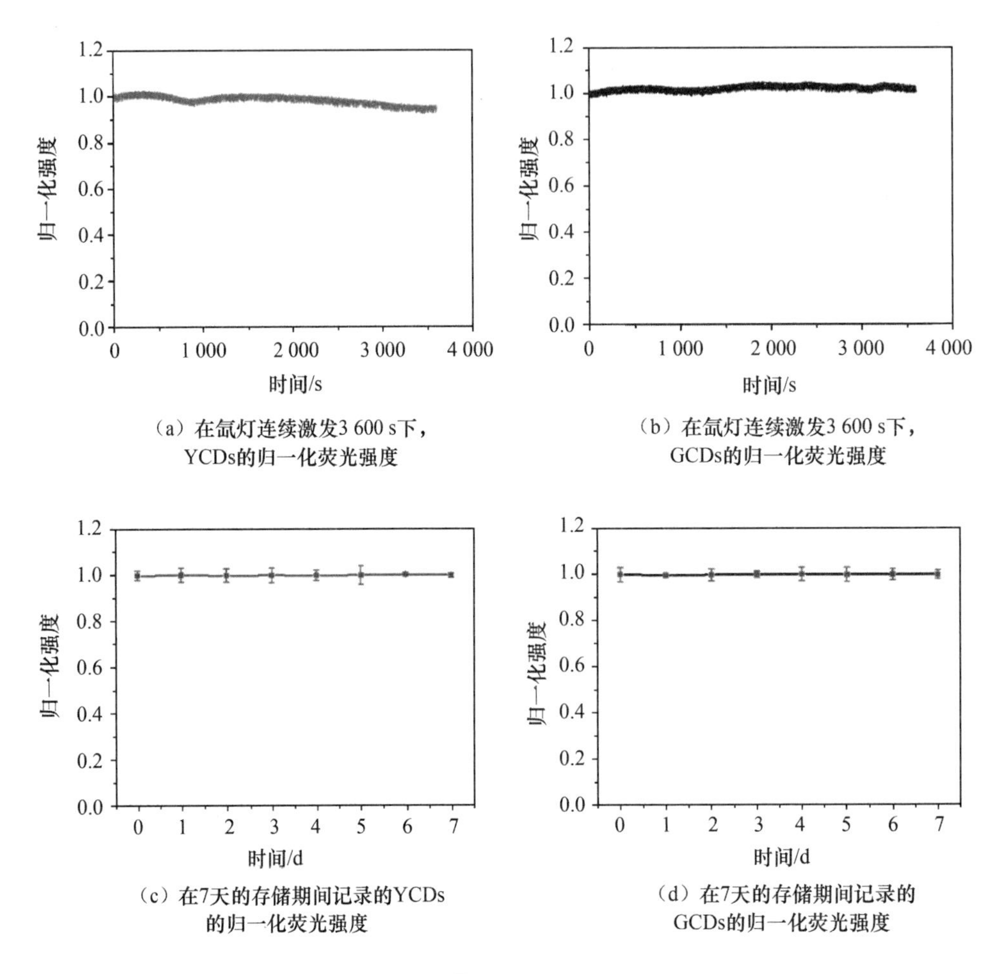

(a)在氙灯连续激发3 600 s下，YCDs的归一化荧光强度

(b)在氙灯连续激发3 600 s下，GCDs的归一化荧光强度

(c)在7天的存储期间记录的YCDs的归一化荧光强度

(d)在7天的存储期间记录的GCDs的归一化荧光强度

图3－12

3.3.3 YCDs和GCDs的荧光发射机理

通常，荧光材料的发射波长与激发态和基态之间的带隙有关。同时，电子态的能级由荧光材料的固有电子结构决定。为了估算两种CDs的HOMO和LUMO能级，本研究采用循环伏安法测试了CDs的电化学性质，并计算了HOMO和LUMO能级。

从图3－13(a)和图3－13(b)可知，YCDs和GCDs的E_{red}分别为－0.45 V和－0.47 V。从YCDs和GCDs的紫外漫反射光谱分别计算出光学带隙E_g为2.39 eV和2.61 eV[图3－13(c)和图3－13(d)]。因此，计算得到YCDs和GCDs的E_{LUMO}分别为－3.95 eV和－3.93 eV，E_{HOMO}分别为－6.34 eV和－6.54 eV。由此可以看出，YCDs和

GCDs 具有不同的 E_{LUMO} 和 E_{HOMO}。同时，当 CDs 的荧光发射红移时，相应的 E_g 随 HOMO 水平的增加和 LUMO 水平的减少而减小。

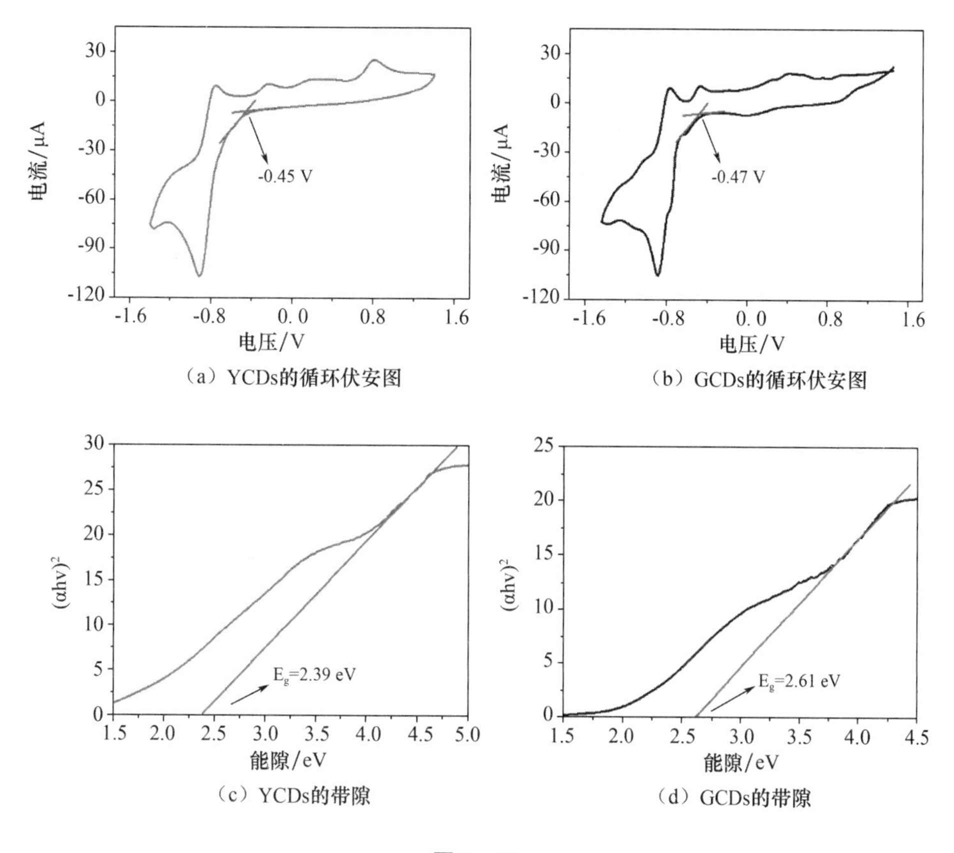

（a）YCDs的循环伏安图

（b）GCDs的循环伏安图

（c）YCDs的带隙

（d）GCDs的带隙

图 3－13

CDs 的结构是如何引起能级变化并影响荧光发射是值得研究的问题。迄今为止，有两种解释多色 CDs 的荧光发射的机理已被广泛接受。一种是基于量子限域效应由碳核大小决定的荧光发射机理[50]。另一个与表面状态有关，这取决于表面官能团[33,41]。在这项研究中，通过 TEM 和 AFM 的结果发现，两个样品均显示出相似的粒径和较宽的粒径分布，这表明量子限域效应不能成为决定两种 CDs 的荧光发射的主要因素。结构表征表明，两种 CDs 具有不同的表面官能团。因此，表面态被认为是控制荧光发射的主要因素，而非量子限域效应。此外，两种 CDs 的荧光的 pH 依赖性进一步证实了表面态是多色荧光发射的主要原因[41]。事实上，与氧氮相关的官能团通常会影响 CDs 的荧光发射[44,51,52]。从表 3－1 中可以看出 YCDs 比 GCDs 含有更多的－OH 基团，－OH 含量的增加将引起 CDs 带隙的减小，减小的带隙最终会引起荧光发射峰的红移[53]。同时，与 GCDs 相比，$-NO_2$ 基团出现在 YCDs 的表面上。据报道，CDs 表面的 $-NO_2$ 基团可以改变电子结构进而影响其能级,导致 CDs 的荧光发射发生红移[44]。因此,两种 CDs 的荧光发射与它们表面上的 - OH 和 - NO_2 基团的含量密切相关。也就是说,YCDs 中丰富的 - OH 和 - NO_2 基团调控了 HOMO 和 LUMO 能量,引起了荧光发射发生红移(图 3 - 14)。

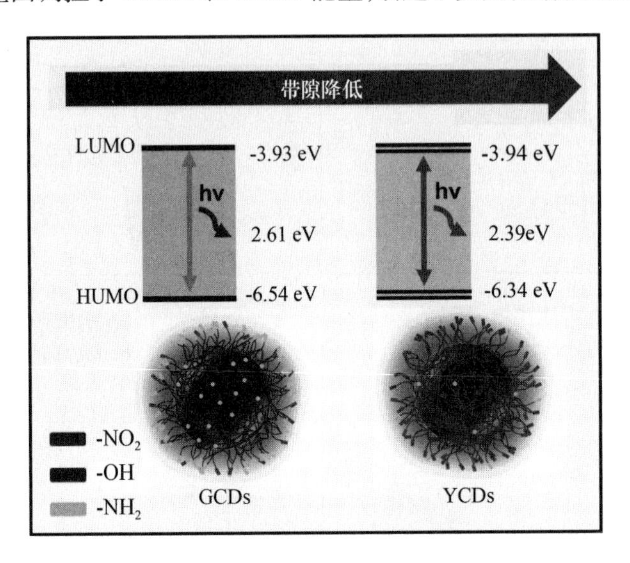

图 3 - 14　YCDs 和 GCDs 的能级图

3.3.4　基于 YCDs 的对硝基苯酚荧光检测

为了评估 YCDs 的应用价值,本工作将 YCDs 作为荧光探针用于 p - NP 的定量分析。如图 3 - 15 所示,为了获得 YCDs 最佳的传感性能,实验优化了检测的 pH 值条件。当传感系统的 pH 值达到 9.0 时,荧光淬灭效率达到最大值。因此,随后的实验在 pH 值为 9.0 下进行。

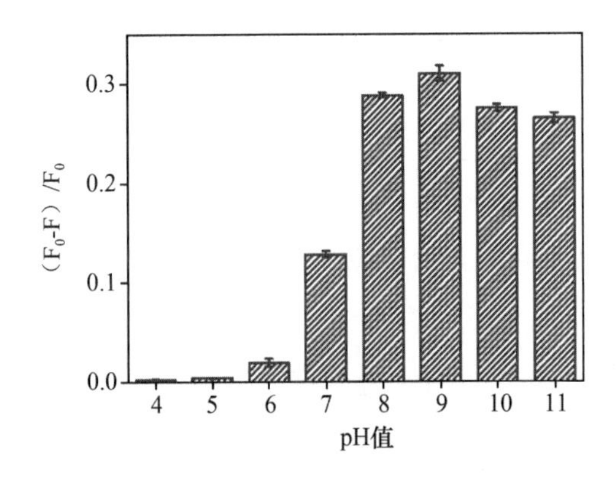

图 3 - 15　在不同 pH 值下,YCDs 对 p - NP 的荧光响应

(YCDs 的浓度 10 μg L^{-1},p - NP 的浓度 25 μmol · L^{-1})

从图3－16(a)可知,在最佳pH值条件下,向YCDs溶液中添加一系列不同浓度(0～140 μmol·L^{-1})p－NP后,YCDs的荧光强度逐渐降低。图3－16(b)展示了淬灭率$(F_0-F)/F_0$与p－NP浓度之间的关系。当p－NP浓度在0至140 μmol·L^{-1}时,$(F_0-F)/F_0$与p－NP浓度之间为正相关关系。当p－NP浓度在0.2至50 μmol·L^{-1}的范围内时,$(F_0-F)/F_0$与p－NP浓度之间展现了良好的线性关系,线性回归方程为:$(F_0-F)/F_0=0.012\ 3C+0.019\ 5$,($R^2=0.997\ 8$)。信噪比为3时,检测限低至0.08 μmol·$L^{-1}$,低于美国环境保护局规定的饮用水中p－NP的最大允许含量(0.43 μmol·L^{-1})[15]。此外,该方法的检测性能可以与之前报道的用于p－NP检测的方法相媲美(表3－2)。

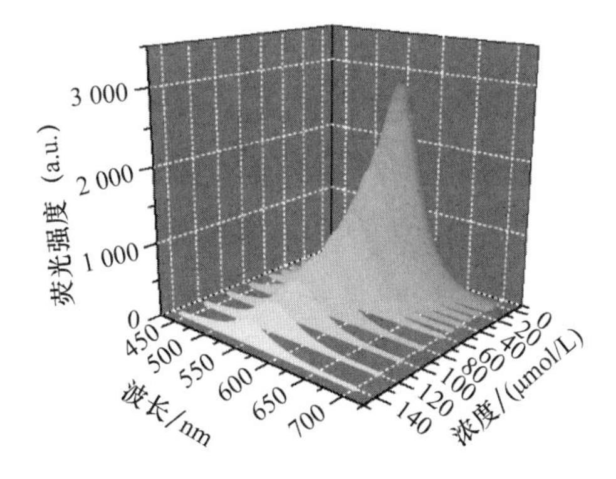

(a)添加不同浓度的p-NP后YCDs的荧光光谱

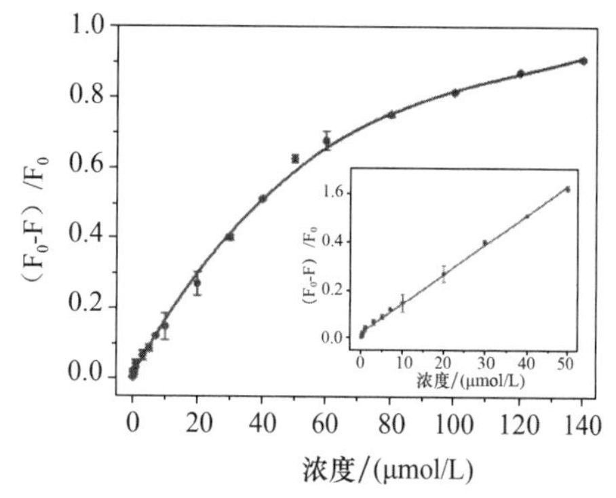

(b)荧光淬灭率$(F_0-F)/F_0$与p-NP浓度之间的关系[插图为$(F_0-F)/F_0$与p-NP浓度之间线性关系]

图3－16

表3－2　检测p－NP分析方法的比较

分析方法	线性范围/(μmol·L^{-1})	检出限/(μmol·L^{-1})	参考文献
电化学法	1－300	0.6	[54]
电化学法	74－375	2.03	[55]
电化学法	20－2400	0.3	[56]
电化学法	5－310	0.183	[57]
荧光法	1－40	0.34	[58]
荧光法	0－12	0.15	[59]
荧光法	0.3－30	0.11	[60]
荧光法	0.5－60	0.26	[61]
荧光法	0.2－50	0.08	本工作

随后,评估了 YCDs 传感器对 p – NP 的选择性(图 3 – 17)。几种结构相似的芳香族化合物,例如邻硝基苯酚(o – NP)、间硝基苯酚(m – NP)、对甲酚(p – MP)、邻甲酚(o – MP)、间甲酚(m – MP)、甲苯(PhMe)、苯胺(AN)、间苯三酚(PG)、对氯苯酚(p – CP)和对苯二胺(p – PD)在相同条件下对 YCDs 荧光的影响可忽略不计。此外,从图 3 – 17 中可以看出,包括 Ag^{+}、Al^{3+}、Ba^{2+}、Ca^{2+}、Cd^{2+}、Co^{2+}、Cr^{3+}、Cu^{2+}、Hg^{2+}、K^{+}、Mg^{2+}、Mn^{2+}、Na^{+}、Ni^{2+}、Pb^{2+}和 Zn^{2+} 等一系列金属阳离子不会引起 YCDs 的荧光光谱发生显著变化。这些结果说明 YCDs 不仅表现了对 p – NP 灵敏性,而且对 p – NP 具有较强的选择性。为了评估传感系统的实用性,以 YCDs 作为荧光探针测定了沈阳市新开河和南湖中的 p – NP含量。如表 3 – 3 所示,结果表明在实际样品中未发现 p – NP。使用加标回收法,回收率在 99.4% ~101.6% 之间。这说明所提出的方法具有较高的分析精度和可信度。

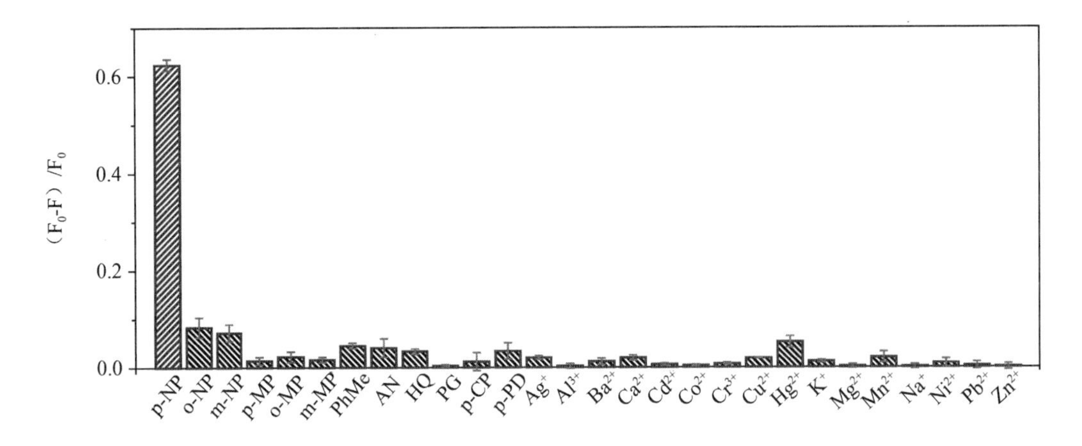

图 3 – 17　p – NP(50 μmol · L^{-1})、芳香族化合物(50 μmol · L^{-1})和常见无机离子(50 μmol · L^{-1})对 YCDs 的荧光淬灭率

表 3 – 3　实际样品中 p – NP 的测定结果

样品名称	加标量/(μmo · L^{-1})	回收量/(μmo · L^{-1})	回收率/%
湖水	0	ND	—
	5	5.07 ± 0.30	101.4
	15	14.94 ± 0.37	99.6
	30	29.95 ± 0.13	99.8
河水	0	ND	—
	5	5.08 ± 0.32	101.6
	15	15.16 ± 0.49	101.1
	30	29.83 ± 1.38	99.4

本工作进一步使用物理化学方法研究了 p－NP 对 YCDs 荧光淬灭的机理。通常情况下，静态淬火、动态淬火和内滤效应都会引起 CDs 的荧光淬火。在这些荧光淬灭机理中，静态淬灭是通过荧光团与淬灭剂之间的强相互作用形成基态配合物，从而导致荧光团的荧光淬灭，这种现象通常会改变吸收光谱[62]。在此情况下，测试了 YCDs、p－NP 以及 YCDs 和 p－NP 混合物的吸收光谱。如图 3－18(a)所示，在 p－NP 存在下 YCDs 的实验吸收光谱与理论上的光谱完全重叠。该结果表明，YCDs 和 p－NP 之间的相互作用极弱，并且在它们之间不形成络合物。因此，排除了静态淬火的可能性。另一方面，动态淬灭机理会使荧光团的荧光寿命发生了变化[63]。在此，YCDs 的荧光寿命约为 2.48 ns。添加 p－NP 后，YCDs 的荧光寿命为 2.42 ns，变化很小[图 3－18(b)]。因此，动态淬灭也可以被排除。通常情况下，有效的内滤效应需要淬灭剂的吸收带与荧光团的激发带和/或发射带之间有尽可能多的光谱重叠[64]。如图 3－18(c)所示，随着 pH 值

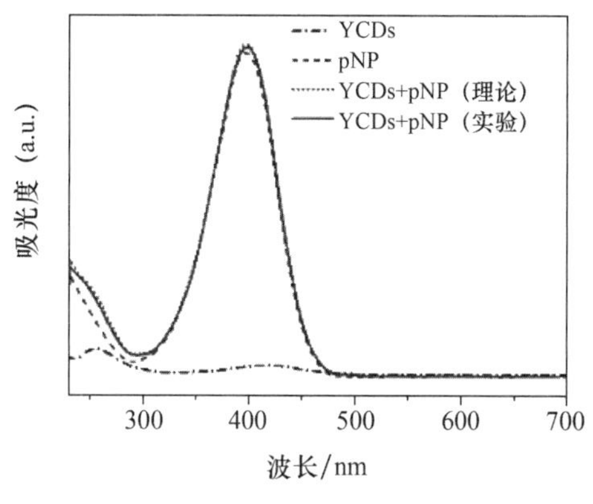

（a）YCDs，p-NP和YCDs/p-NP（理论值）和YCDs/p-NP（实验值）的UV-vis吸收光谱

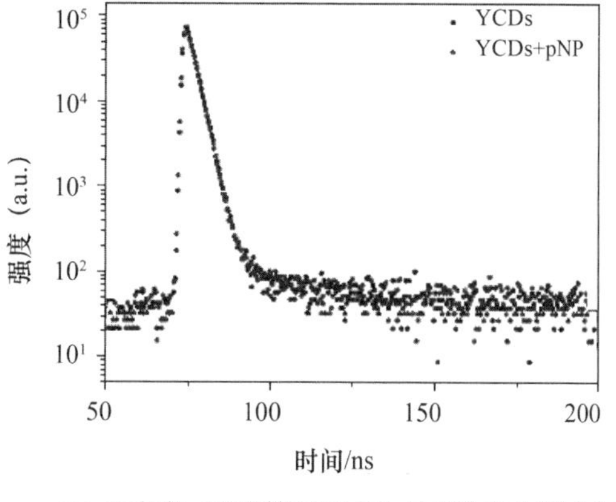

（b）不存在p-NP的情况下YCDs的荧光寿命衰减曲线

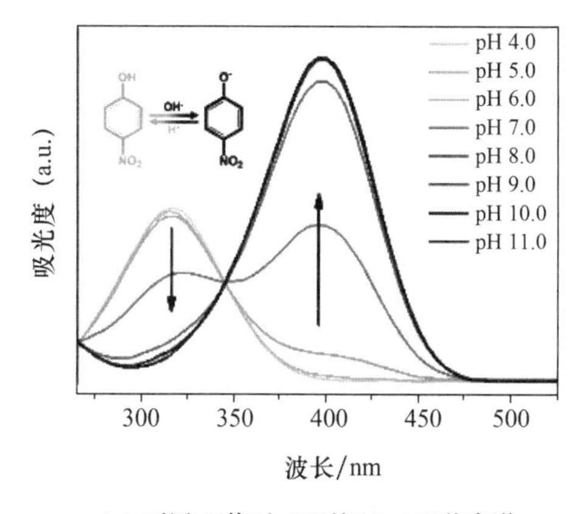

（c）不同pH值下p-NP的UV-vis吸收光谱

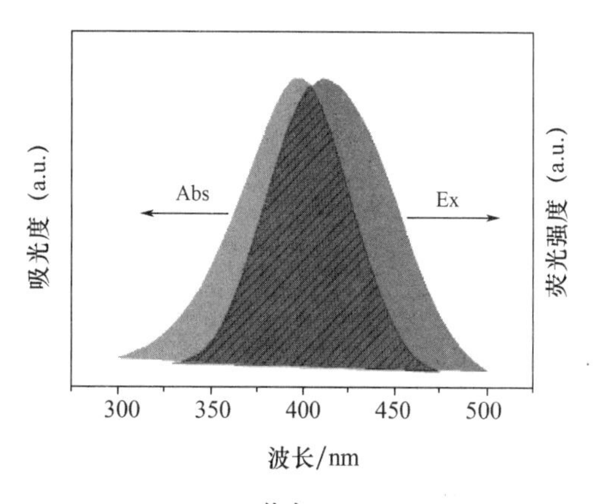

（d）p-NP（pH值为9.0）的UV-vis吸收光谱和YCDs的荧光激发光谱

图 3－18

的增加,p－NP 吸收峰在约 320 至 400 nm 的范围内发生红移。红移的发生是因为碱性条件下 p－NP 氧阴离子的形成[65]。当 pH 值为 9.0 时,YCDs 的荧光激发光谱几乎与 p－NP 的吸收光谱重叠[图 3－18(d)]。由此可见,尽管内滤效应对 YCDs 的荧光强度具有很大的影响,但是由于淬灭没有涉及的激发态能量/电子转移,因此荧光寿命保持不变。同时,在内滤效应中,CDs 和淬灭剂之间没有直接的相互作用。因此,该检测系统的荧光淬灭机理可确定为内滤效应。此外,考虑到 GCDs 与 YCDs 的同源性,本工作在相同条件下将 GCDs 用于分析 p－NP。由于 GCDs 的荧光激发光谱和 p－NP 的吸收光谱的重叠较小,因此 p－NP 对 GCDs 的荧光淬灭作用低于 YCDs(图 3－19)。

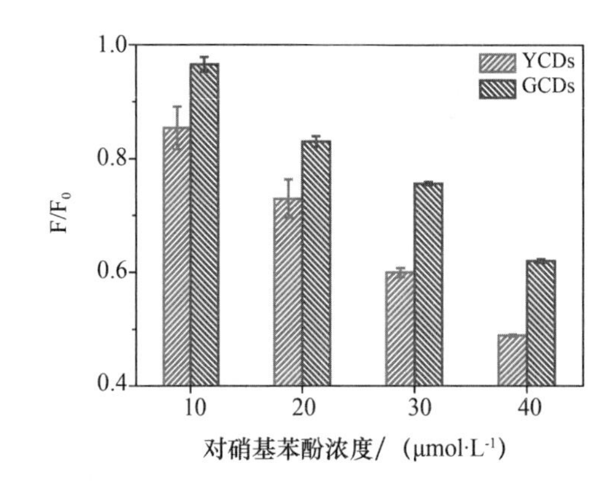

图 3－19　两种 CDs 归一化荧光强度与 p－NP 浓度之间关系

3.3.5　基于 GCDs 的重水中水含量荧光检测

与在 H_2O 中一样,GCDs 粉末可以均匀分散在 D_2O 中形成透明溶液。出人意料的是,GCDs 在 H_2O 和 D_2O 中展现了不同的荧光性能。如图 3－20(a)所示,分别测试了分散在 H_2O 和 D_2O 中 GCDs 的荧光衰减曲线。GCDs 在 H_2O 和 D_2O 中的平均寿命分别为 1.75 ns 和 2.11 ns。可以看出,D_2O 中 GCDs 的寿命比 H_2O 中的寿命长,这表明 H_2O 对 GCDs 的激发态具有比 D_2O 更强的淬灭作用。该现象主要是由 O－H 振荡的淬灭效果高于 O－D 振荡的淬灭效果所致[19]。之后,本工作研究了 D_2O 中不同含量的 H_2O 对 YCDs 的荧光强度的影响。如图 3－20(b)所示,GCDs 的荧光强度随着 D_2O 中 H_2O 含量的增加而降低。此外,荧光强度比 F_0/F 与 D_2O 中 H_2O 含量成正比,范围为体积分数 0.5～40[图 3－20(c)],并且可以获得良好的线性关系,线性方程为:$F_0/F=0.013C+0.999$,($R^2=0.992\ 2$),信噪比为 3 时,检出限为体积分数 0.17。

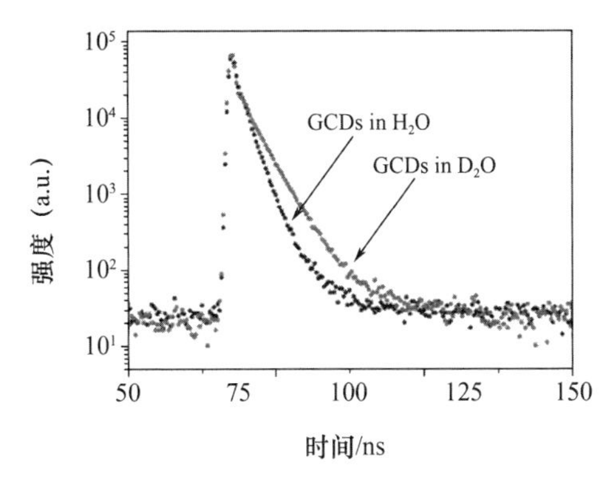

（a）GCDs在H_2O和D_2O中的荧光寿命衰减曲线

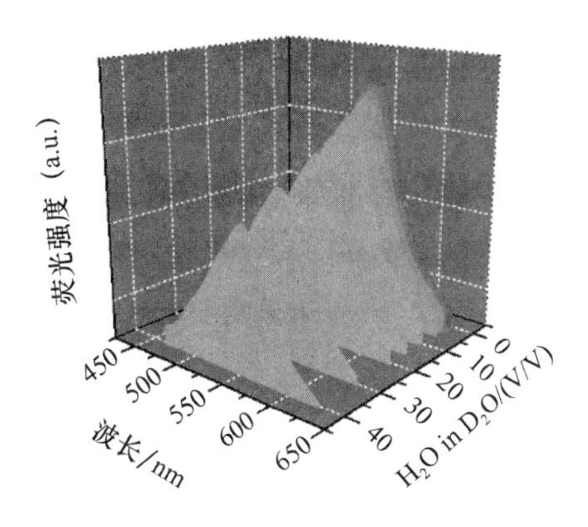

（b）添加不同含量的H_2O（体积分数为0~40）的D_2O中GCDs的荧光光谱变化

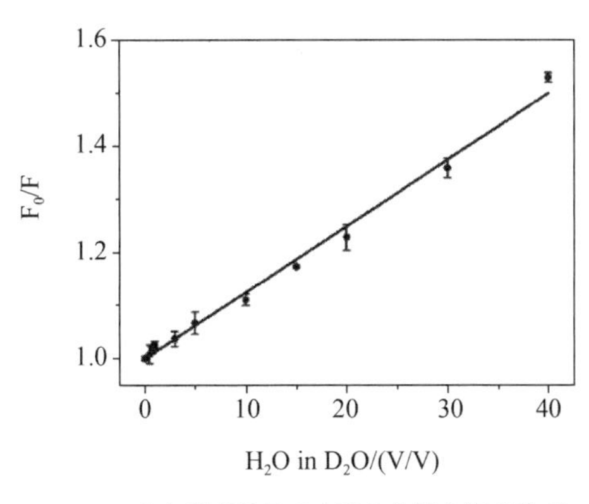

（c）F_0/F与D_2O中H_2O含量之间的关系

图 3 - 20

考虑到测试环境的复杂性，研究了 GCDs 抗干扰能力（图 3 - 21）。结果表明，GCDs 对潜在的干扰物具有极强的抗干扰性。因此，证明了 GCDs 具有作为纳米探针检测 D_2O 中 H_2O 含量的潜力。此外，通过加标回收法测定市售 D_2O 产品中 H_2O 含量用于检验 GCDs 作为荧光探针的实际应用能力。加入 H_2O 的量在体积分数 5 ~ 30 之间，回收率为 99.6% ~ 101.5%（表 3 - 4）。这些结果表明，所提出的方法具有较高的分析精度和可信度。此外，基于 YCDs 与 GCDs 的同源性，本工作在相同条件下将 YCDs 用于分析 D_2O 中 H_2O 的含量。结果显示，由于不同表面态因素的影响，YCDs 表现出了低于 GCDs 区分 H_2O 和 D_2O 的能力（图 3 - 22）。

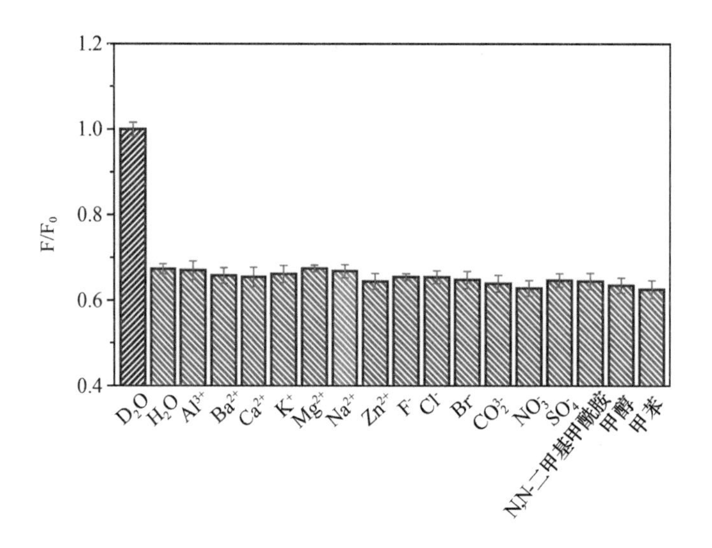

图 3-21 添加体积分数 40 的 H_2O 和添加体积分数 40 含有干扰物的 H_2O 后，GCDs 的 D_2O 溶液荧光响应（干扰物浓度 = 1000 μmol · L^{-1}）

表 3-4 实际样品中 H_2O 的测定结果

样品编号	加标量/(μmo · L^{-1})	回收量/(μmo · L^{-1})	回收率/%
1	0	ND	—
2	5	4.98 ± 0.16	99.6
3	15	14.97 ± 0.22	99.8
4	30	30.44 ± 0.19	101.5

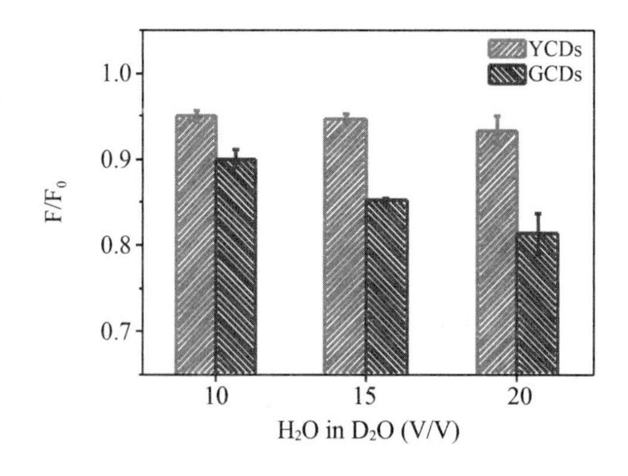

图 3-22 两种 CDs 归一化荧光强度与 D_2O 中 H_2O 体积分数之间关系

3.4　小结

本研究采用室温氧化融合法，以 o – PD 和 HQ 为前驱体，绿色制备了不同表面态 o – PD 衍生 CDs。此方法不同于需要高温才能使有机小分子热解或碳化的传统制备 CDs 的手段，而是通过室温下自发的氧化/聚合反应制备的 CDs。这不仅使 CDs 本身成为绿色材料，而且使制备过程成为绿色方法。基于不同的物理化学性质，通过硅胶色谱柱可以将粗制 CDs 纯化为两种具有不同荧光发射的 CDs（YCDs 和 GCDs）。研究发现，两种 CDs 的荧光发射与 – OH 和 – NO_2 的含量调控的不同表面态有关。最后，基于荧光内滤效应，YCDs 展现出了对 p – NP 良好的选择性。线性范围为 0.2 ~ 50 μmol · L^{-1}，检出限为 0.08 μmol · L^{-1}。将 YCDs 用于实际样品中 p – NP 定量分析，结果令人满意。此外，受表面大量 – NH_2 与 O – H/O – D 相互作用的影响，GCDs 可以作为荧光探针用于检测 D_2O 中 H_2O 的含量，线性范围为体积分数 0.5 ~ 40，检出限为体积分数 0.17。本工作认为采用室温氧化融合法制备不同表面态 CDs 将极大地促进绿色制备多色/多用途 CDs 的方法的进步。

虽然，本工作通过合理选择前驱体，将制备不同表面态的 CDs 的制备条件由加热反应改为了室温反应，使制备 CDs 的条件更加绿色化。然而，制备出的 CDs 的荧光发射范围还是有一定的局限性，这将限制 CDs 在更多领域的应用。因此，在下一章的工作中，希望开发一种在室温下制备更长发射波长 o – PD 衍生 CDs 的方法。

参考文献

[1] Qiao G, Chen G, Wen Q, et al. Rapid conversion from common precursors to carbon dots in large scale: Spectral controls, optical sensing, cellular imaging and LEDs application [J]. Journal of Colloid and Interface Science, 2020, 580: 88 – 98.

[2] Liu M L, Chen B B, Li C M, et al. Carbon dots: Synthesis, formation mechanism, fluorescence origin and sensing applications [J]. Green Chemistry, 2019, 21 (3): 449 – 471.

[3] Zhang Z, Zhang D, Shi C, et al. 3,4 – Hydroxypyridinone – modified carbon quantum dot as a highly sensitive and selective fluorescent probe for the rapid detection of uranyl ions [J]. Environmental Science: Nano, 2019, 6 (5): 1457 – 1465.

[4]Zhao S, Wu S, Jia Q, et al. Lysosome – targetable carbon dots for highly efficient photothermal/photodynamic synergistic cancer therapy and photoacoustic/two – photon excited fluorescence imaging [J]. Chemical Engineering Journal, 2020, 388: 124212.

[5]Liu X, Li T, Hou Y, et al. Microwave synthesis of carbon dots with multi – response using denatured proteins as carbon source [J]. RSC Advances, 2016, 6 (14): 11711 – 11718.

[6]Gao X, Du C, Zhuang Z, et al. Carbon quantum dot – based nanoprobes for metal ion detection [J]. Journal of Materials Chemistry C, 2016, 4 (29): 6927 – 6945.

[7]Krysmann M J, Kelarakis A, Giannelis E P. Photoluminescent carbogenic nanoparticles directly derived from crude biomass [J]. Green Chemistry, 2012, 14 (11): 3141.

[8]Ding Z, Li F, Wen J, et al. Gram – scale synthesis of single – crystalline graphene quantum dots derived from lignin biomass [J]. Green Chemistry, 2018, 20 (6): 1383 – 1390.

[9]Qiao Z – A, Wang Y, Gao Y, et al. Commercially activated carbon as the source for producing multicolor photoluminescent carbon dots by chemical oxidation [J]. Chemical Communications, 2010, 46 (46): 8812 – 8814.

[10]Li Y, Zhong X, Rider A E, et al. Fast, energy – efficient synthesis of luminescent carbon quantum dots [J]. Green Chemistry, 2014, 16 (5): 2566 – 2570.

[11]Peng H, Travas – Sejdic J. Simple aqueous solution route to luminescent carbogenic dots from carbohydrates [J]. Chemistry of Materials, 2009, 21 (23): 5563 – 5565.

[12]Xiao S J, Chu Z J, Zuo J, et al. Fluorescent carbon dots: Facile synthesis at room temperature and its application for Fe^{2+} sensing [J]. Journal of Nanoparticle Research, 2017, 19 (2).

[13]Liu H, Ye T, Mao C. Fluorescent carbon nanoparticles derived from candle soot [J]. Angewandte Chemie International Edition, 2007, 46 (34): 6473 – 6475.

[14]Yue W, Chen M, Cheng Z, et al. Bioaugmentation of strain Methylobacterium sp. C1 towards p – nitrophenol removal with broad spectrum coaggregating bacteria in sequencing batch biofilm reactors [J]. Journal of Hazardous Materials, 2018, 344: 431 – 440.

[15]Hu Y, Gao Z. Sewage sludge in microwave oven: A sustainable synthetic approach toward carbon dots for fluorescent sensing of para – nitrophenol [J]. Journal of Hazardous

Materials, 2020, 382: 121048.

[16] Ma X, Wu Y, Devaramani S, et al. Preparation of GO - COOH/AuNPs/ZnAPTPP nanocomposites based on the π - π conjugation: Efficient interface for low - potential photoelectrochemical sensing of 4 - nitrophenol [J]. Talanta, 2018, 178: 962 - 969.

[17] Vilian A T E, Choe S R, Giribabu K, et al. Pd nanospheres decorated reduced graphene oxide with multi - functions: Highly efficient catalytic reduction and ultrasensitive sensing of hazardous 4 - nitrophenol pollutant [J]. Journal of Hazardous Materials, 2017, 333: 54 - 62.

[18] Deng P, Xu Z, Feng Y, et al. Electrocatalytic reduction and determination of p - nitrophenol on acetylene black paste electrode coated with salicylaldehyde - modified chitosan [J]. Sensors and Actuators B: Chemical, 2012, 168: 381 - 389.

[19] Luo Y, Li C, Zhu W, et al. A facile strategy for the construction of purely organic optical sensors capable of distinguishing D_2O from H_2O [J]. Angewandte Chemie International Edition, 2019, 58 (19): 6280 - 6284.

[20] Xia J, Yu Y - L, Wang J - H. Fe^{3+} - Catalyzed low - temperature preparation of multicolor carbon polymer dots with the capability of distinguishing D_2O from H_2O [J]. Chemical Communications, 2019, 55 (83): 12467 - 12470.

[21] Zheng F, Luo Y, Li C, et al. A water - soluble sensor for distinguishing D_2O from H_2O by dual - channel absorption/fluorescence ratiometry [J]. Chemical Communications, 2022, 58 (92): 12863 - 12866.

[22] Mir I A, Rawat K, Bohidar H B. CuInGaSe nanocrystals for detection of trace amount of water in D_2O (at ppm level) [J]. Crystal Research and Technology, 2016, 51 (10): 561 - 568.

[23] Dunning S G, Nuñez A J, Moore M D, et al. A sensor for trace H_2O detection in D_2O [J]. Chem, 2017, 2 (4): 579 - 589.

[24] Li H, He X, Liu Y, et al. One - step ultrasonic synthesis of water - soluble carbon nanoparticles with excellent photoluminescent properties [J]. Carbon, 2011, 49 (2): 605 - 609.

[25] Liu S, Qin X, Tian J, et al. Photochemical preparation of fluorescent 2,3 - diaminophenazine nanoparticles for sensitive and selective detection of Hg(II) ions [J]. Sensors and Actuators B: Chemical, 2012, 171 - 172: 886 - 890.

[26] Huang T, Wu T, Zhu Z, et al. Self – assembly facilitated and visible light – driven generation of carbon dots [J]. Chemical Communications, 2018, 54 (47): 5960 – 5963.

[27] Chen B B, Liu Z X, Deng W C, et al. A large – scale synthesis of photoluminescent carbon quantum dots: A self – exothermic reaction driving the formation of the nanocrystalline core at room temperature [J]. Green Chemistry, 2016, 18 (19): 5127 – 5132.

[28] Yu H, Dong F, Li B, et al. Co(Ⅱ) triggered radical reaction between SO_2 and o – phenylenediamine for highly selective visual colorimetric detection of SO_2 gas and its derivatives [J]. Sensors and Actuators B: Chemical, 2019, 299: 126983.

[29] Li T, E S, Wang J, et al. Regulating the properties of carbon dots via a solvent – involved molecule fusion strategy for improved sensing selectivity [J]. Analytica Chimica Acta, 2019, 1088: 107 – 115.

[30] Shen D, Long Y, Wang J, et al. Tuning the fluorescence performance of carbon dots with a reduction pathway [J]. Nanoscale, 2019, 11 (13): 5998 – 6003.

[31] Liu X, Li T, Wu Q, et al. Carbon nanodots as a fluorescence sensor for rapid and sensitive detection of Cr(Ⅵ) and their multifunctional applications [J]. Talanta, 2017, 165: 216 – 222.

[32] Kozák O, Sudolská M, Pramanik G, et al. Photoluminescent carbon nanostructures [J]. Chemistry of Materials, 2016, 28 (12): 4085 – 4128.

[33] Ding H, Yu S – B, Wei J – S, et al. Full – color light – emitting carbon dots with a surface – state – controlled luminescence mechanism [J]. ACS Nano, 2016, 10 (1): 484 – 491.

[34] Lyu B, Li H – J, Xue F, et al. Facile, gram – scale and eco – friendly synthesis of multi – color graphene quantum dots by thermal – driven advanced oxidation process [J]. Chemical Engineering Journal, 2020, 388: 124285.

[35] Liu M – X, Chen S, Ding N, et al. A carbon – based polymer dot sensor for breast cancer detection using peripheral blood immunocytes [J]. Chemical Communications, 2020, 56 (20): 3050 – 3053.

[36] Zhang Y, Yuan R, He M, et al. Multicolour nitrogen – doped carbon dots: Tunable photoluminescence and sandwich fluorescent glass – based light – emitting diodes [J]. Nanoscale, 2017, 9 (45): 17849 – 17858.

[37] Nie H, Li M, Li Q, et al. Carbon dots with continuously tunable full – color emission

and their application in ratiometric pH sensing [J]. Chemistry of Materials, 2014, 26 (10): 3104 - 3112.

[38] Yarur F, Macairan J - R, Naccache R. Ratiometric detection of heavy metal ions using fluorescent carbon dots [J]. Environmental Science: Nano, 2019, 6 (4): 1121 - 1130.

[39] Yao K, Lv X, Zheng G, et al. Effects of carbon quantum dots on aquatic environments: Comparison of toxicity to organisms at different trophic levels [J]. Environmental Science & Technology, 2018, 52 (24): 14445 - 14451.

[40] Sun S, Zhang L, Jiang K, et al. Toward high - efficient red emissive carbon dots: Facile preparation, unique properties, and applications as multifunctional theranostic agents [J]. Chemistry of Materials, 2016, 28 (23): 8659 - 8668.

[41] Han L, Liu S G, Dong J X, et al. Facile synthesis of multicolor photoluminescent polymer carbon dots with surface - state energy gap - controlled emission [J]. Journal of Materials Chemistry C, 2017, 5 (41): 10785 - 10793.

[42] Dong Y, Li J, Zhang L. 3D hierarchical hollow microrod via in - situ growth 2D SnS nanoplates on MOF derived Co, N co - doped carbon rod for electrochemical sensing [J]. Sensors and Actuators B: Chemical, 2020, 303: 127208.

[43] Shu Y, Lu J, Mao Q - X, et al. Ionic liquid mediated organophilic carbon dots for drug delivery and bioimaging [J]. Carbon, 2017, 114: 324 - 333.

[44] Zhan J, Geng B, Wu K, et al. A solvent - engineered molecule fusion strategy for rational synthesis of carbon quantum dots with multicolor bandgap fluorescence [J]. Carbon, 2018, 130: 153 - 163.

[45] Chen Y H, Chen G, Lee D J. Synthesis of low surface energy thin film of polyepichlorohydrin - triazole - ols [J]. Journal of Colloid and Interface Science, 2020, 575: 452 - 463.

[46] Saini B, Singh R R, Nayak D, et al. Biocompatible pH - responsive luminescent coacervate nanodroplets from carbon dots and poly (diallyldimethylammonium chloride) toward theranostic applications [J]. ACS Applied Nano Materials, 2020, 3 (6): 5826 - 5837.

[47] Yuan B, Guan S, Sun X, et al. Highly efficient carbon dots with reversibly switchable green - red emissions for trichromatic white light - emitting diodes [J]. ACS Applied Materials & Interfaces, 2018, 10 (18): 16005 - 16014.

[48] Zhang T, Zhu J, Zhai Y, et al. A novel mechanism for red emission carbon dots: Hydrogen bond dominated molecular states emission [J]. Nanoscale, 2017, 9 (35): 13042 - 13051.

[49] Hu S, Trinchi A, Atkin P, et al. Tunable photoluminescence across the entire visible spectrum from carbon dots excited by white light [J]. Angewandte Chemie International Edition, 2015, 54 (10): 2970 - 2974.

[50] Tian Z, Zhang X, Li D, et al. Full - color inorganic carbon dot phosphors for white - light - emitting diodes [J]. Advanced Optical Material, 2017, 5: 1700416.

[51] Ding H, Wei J - S, Zhang P, et al. Solvent - controlled synthesis of highly luminescent carbon dots with a wide color gamut and narrowed emission peak widths [J]. Small, 2018, 14 (22): 1800612.

[52] Hola K, Sudolska M, Kalytchuk S, et al. Graphitic nitrogen triggers red fluorescence in carbon dots [J]. ACS Nano, 2017, 11 (12): 12402 - 12410.

[53] Sk M A, Ananthanarayanan A, Huang L, et al. Revealing the tunable photoluminescence properties of graphene quantum dots [J]. Journal of Materials Chemistry C, 2014, 2 (34): 6954 - 6960.

[54] Yin H, Zhou Y, Ai S, et al. Electrochemical oxidative determination of 4 - nitrophenol based on a glassy carbon electrode modified with a hydroxyapatite nanopowder [J]. Microchimica Acta, 2010, 169 (1 - 2): 87 - 92.

[55] Khan S B, Akhtar K, Bakhsh E M, et al. Electrochemical detection and catalytic removal of 4 - nitrophenol using CeO_2 - Cu_2O and CeO_2 - Cu_2O/CH nanocomposites [J]. Applied Surface Science, 2019, 492: 726 - 735.

[56] Noorbakhsh A, Mirkalaei M M, Yousefi M H, et al. Electrodeposition of cobalt oxide nanostructure on the glassy carbon electrode for electrocatalytic determination of para - nitrophenol [J]. Electroanalysis, 2014, 26 (12): 2716 - 2726.

[57] Wang M, Liu Y, Yang L, et al. Bimetallic metal - organic framework derived FeO / TiO_2 embedded in mesoporous carbon nanocomposite for the sensitive electrochemical detection of 4 - nitrophenol [J]. Sensors and Actuators B: Chemical, 2019, 281: 1063 - 1072.

[58] Geng S, Lin S M, Liu S G, et al. A new fluorescent sensor for detecting p - nitrophenol based on β - cyclodextrin - capped ZnO quantum dots [J]. RSC Advances, 2016, 6

(89): 86061 - 86067.

[59] Li W, Zhang H, Chen S, et al. Synthesis of molecularly imprinted carbon dot grafted YVO4:Eu^{3+} for the ratiometric fluorescent determination of paranitrophenol [J]. Biosensors and Bioelectronics, 2016, 86: 706 - 713.

[60] Zhang S, Zhang D, Ding Y, et al. Bacteria - derived fluorescent carbon dots for highly selective detection of p - nitrophenol and bioimaging [J]. Analyst, 2019, 144 (18): 5497 - 5503.

[61] Han L, Liu S G, Liang J Y, et al. pH - mediated reversible fluorescence nanoswitch based on inner filter effect induced fluorescence quenching for selective and visual detection of 4 - nitrophenol [J]. Journal of Hazardous Materials, 2019, 362: 45 - 52.

[62] Sun J, Zhao J, Wang L, et al. Inner filter effect - based sensor for horseradish peroxidase and its application to fluorescence immunoassay [J]. ACS Sensors, 2018, 3 (1): 183 - 190.

[63] Lu S, Wang S, Zhao J, et al. Fluorescence light - up biosensor for microRNA based on the distance - dependent photoinduced electron transfer [J]. Analytical Chemistry, 2017, 89: 8429 - 8436.

[64] Shang L, Dong S. Design of fluorescent assays for cyanide and hydrogen peroxide based on the inner filter effect of metal nanoparticles [J]. Analytical Chemistry, 2009, 81: 1465 - 1470.

[65] Johnson J A, Makis J J, Marvin K A, et al. Size - dependent hydrogenation of p - nitrophenol with Pd nanoparticles synthesized with poly(amido)amine dendrimer templates [J]. The Journal of Physical Chemistry C, 2013, 117 (44): 22644 - 22651.

第4章
氟掺杂的表面态调控法制备不同碳点及其应用

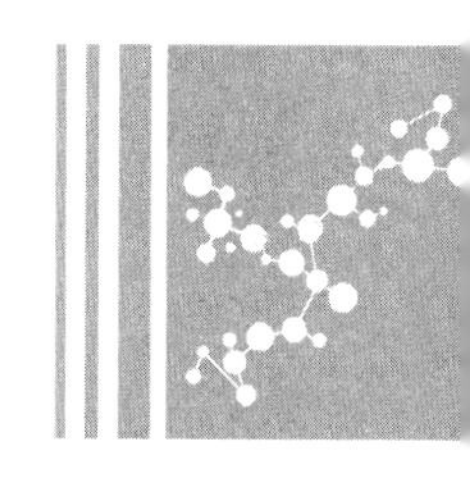

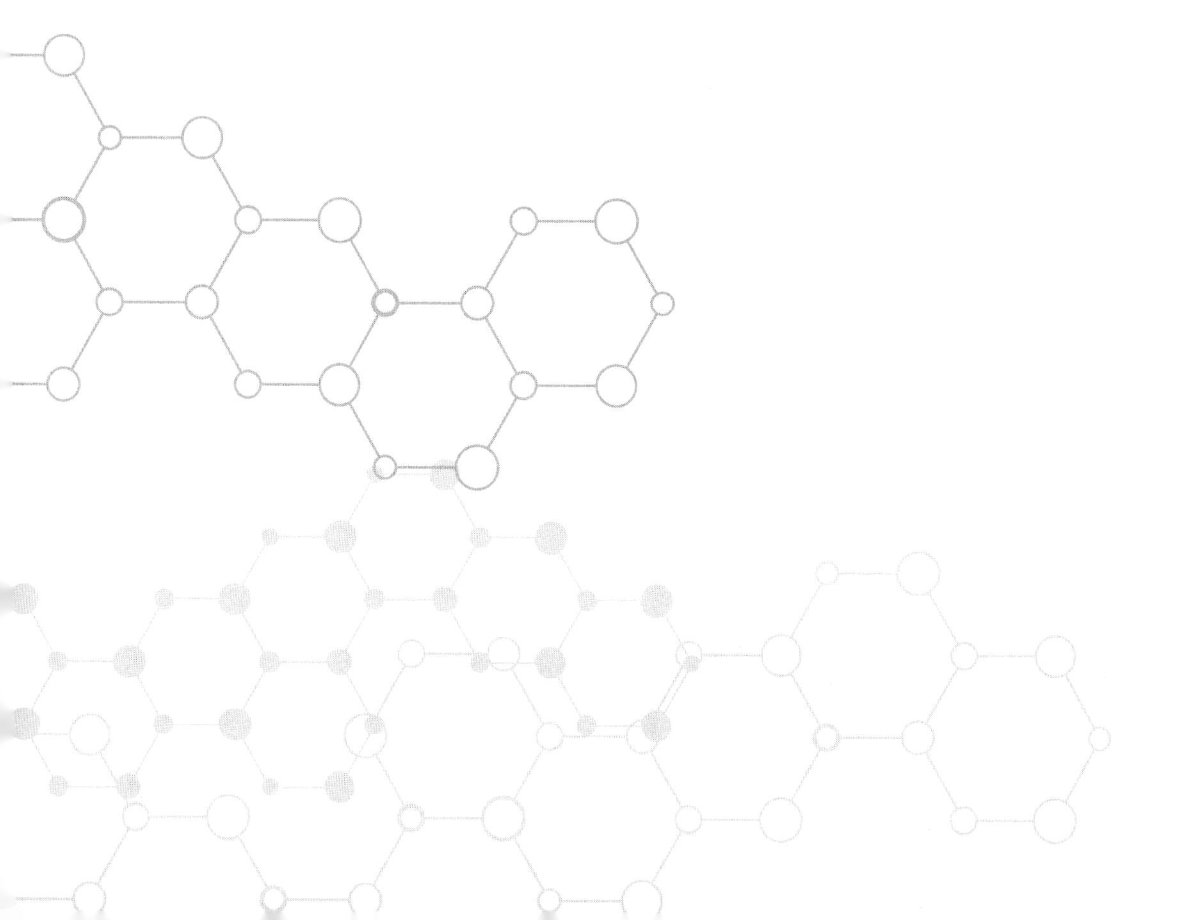

4.1　引言

低维材料是维数比三小的材料，具体来说是二维、一维和零维材料。由于这些材料晶体结构的特异性，因此许多低维度材料展现出了非常奇特的物理现象。近30年来，低维碳纳米材料受到了人们广泛的关注。1985年Smalley等人报道了低维材料的第一位成员，即富勒烯。1991年和2004年发现了另外两种同素异形体，即碳纳米管和石墨烯，引起了科学界的轰动。由于量子限域效应的存在，这些纳米尺度碳表现出许多块状碳材料无可比拟的优异光、电、磁等性质。CDs首次报道于2004年，Scrivens等人在用凝胶电泳纯化电弧放电制得的单壁碳纳米管时偶然发现[1]。2006年，Sun等人通过激光烧蚀－硝酸氧化－有机物钝化的方法制备荧光碳纳米粒子，并首次将其命名为CDs[2]。由于制备CDs的原料来源十分广泛（例如，块状碳材料、各类有机小分子、共轭及非共轭聚合物和生物质材料等），CDs的制备方法也多种多样。CDs作为一种新型的荧光纳米材料，因其具有低毒、优异的光学性能、高化学稳定性、优良的生物相容性和良好的光稳定性等优点，近年来受到广泛关注[3-6]。到目前为止，水热法、溶剂热法和微波法等多种方法已经广泛应用于CDs制备中[7-9]。这些已经报道的方法通常表现出高效性，但这些制备过程的高能量消耗必然伴随着环境污染；同时，烦琐的操作也为后续的规模化生产埋下隐患。常温下制备CDs作为一种低能耗的制备方法，已成为近年来的研究热点。Liu等人报道了通过在室温下混合多巴胺、过氧化氢和o－PD制备CDs，获得的CDs提供了研究细胞极性和与自噬相关的生理或病理过程的潜力[10]。Huang等人在室温下以乙二醛和o－PD为碳源和氮源制备CDs，并将其作为荧光探针用于盐酸四环素的检测[11]。尽管制备CDs的方法已得到迅速发展，但由于缺乏对室温下制备CDs的荧光来源的深刻认识，所以在室温下制备CDs的荧光发射大多是具有随机性的。而荧光作为CDs的固有性质，清楚地认识其结构和荧光之间的关系对于CDs未来的发展十分重要。因此，设计在室温下制备具有所需荧光发射的CDs具有非常重要的意义。基于荧光材料的化学单元对光学特性与表面态/结构进行系统调节是调控荧光材料发光性质的常用方法[12]。通过此方法在高能耗制备CDs领域已经开发了一些协同策略来系统地调整CDs的表面态/结构以精准地控制CDs的光学特性[13]。其中，化学掺杂无疑是改变CDs的荧光特性的一种简单、快速和有效的方法。Fan等人通过溶剂热法制备了含硫CDs，他们证明了硫成分的引入可以显著增加石墨氮的含量从而有效地缩小CDs的带隙，引起荧光发射红移[14]。Dong等人采用一步水热法，以柠檬酸一水合物和半胱氨酸为前驱体制备了氮硫共掺杂 CDs。由于掺杂的氮、硫原子的协同调控作用,制备的 CDs 量子产率达 70% 以上[15]。因此,将化学掺杂合理运用于低能耗制备 CDs 领域调控 CDs 荧光特性是一个不错的选择。

指纹识别技术是一种强有力的工具,在法医学中得到了广泛的应用,因为每个人每根手指的指纹都是独一无二的,不随年龄而变化。在许多情况下,肉眼无法直接看到指纹,因此发展可视化技术是十分必要的。文献中已经建立了设备分析法、化学染色法、粉末除尘法等多种鉴别指纹的方法[16-18]。然而,每种技术都有一些局限性。拉曼光谱、傅里叶变换红外光谱光声成像所需的设备分析技术要昂贵而笨重的仪器,这些在犯罪现场是无法达到[17,18]。化学染色法使用化学试剂如茚三酮、氨基酸和盐类,因此对使用者不可避免的产生有毒和有害影响[19]。粉末除尘技术是在无孔和无孔物体上显影指纹的最常用方法。传统的指纹粉有常规指纹粉和金属指纹粉两种,然而这些方法所用粉末会有非荧光性、对比度低、尺寸不均匀、灵敏度低等特点[20]。与之相比,基于荧光的粉末除尘方法具有较高的灵敏度和空间分辨率[21]。从这些因素出发,发光材料在法医学中提供了可能的应用。将指纹图案给出的复杂排列作为直接生物统计信息已被广泛用于个人识别。在许多情况下,指尖上的汗液/油脂沉积物和下表面之间的光学对比度差,所以用肉眼很难分辨指纹细节,这些指纹称为潜指纹(LFPs)[22,23]。事实上,这种指纹是罪犯遗留现场最常见的证据,它们需要通过某些特殊的技术处理才能清晰可见。因此,设计增强成像方法实现 LFPs 的清晰可视化引起了科研人员极大的兴趣。迄今为止,荧光粉已成为 LFPs 可视化强大的工具,通过荧光粉成像的指纹显示出良好的对比度和准确性[24]。但是,由于荧光粉具有毒性和有害成分,导致传统荧光粉的使用面临巨大的问题。因此,开发一种绿色、高效、成本低廉的荧光材料用于 LFPs 的显现是十分必要的。

钴胺素(CBL),即维生素 B_{12},是自然界中结构最复杂的维生素,也是人体必需的维生素之一,在体内作为生物辅酶参与同型半胱氨酸代谢和甲基丙二酰辅酶 A 与琥珀酰辅酶 A 的转化等多项生理活动。人类作为高等生物,无法自行合成钴胺素,需要通过饮食或药物来补充人体所需的钴胺素。目前,用于治疗的钴胺素形式多样,常见形式有氰钴胺(普通维生素 B_{12})、羟钴胺、甲钴胺和腺苷钴胺。CBL 常见于牛奶、鸡蛋、鸡肉以及沙丁鱼等食物中。CBL 在生长、细胞发育中起着重要作用,人体对 CBL 每天的需求量为 0.40 ~2.80 mg,缺乏 CBL 会导致恶性贫血、神经系统损伤和心脏病等疾病,但过量或不当食用也会导致哮喘、肝病和肾衰竭等疾病[25]。高效液相色谱、化学发光、酶联免疫吸附测定和电化学分析是测定 CBL 含量的常用方法[26]。但是,这些方法都有其自身的局限性，例如：费时、价格昂贵以及需要复杂的样品预处理等。因此，发展一种简单准确而有效的监测技术用于定量分析 CBL 含量是十分必要的。

在本研究中，基于氟掺杂的表面态控制法室温下绿色制备不同表面态 o－PD 衍生 CDs，通过氟掺杂技术实现了 CDs 的荧光调节。通过氟含量对 CDs 表面态的影响，在一定程度上限制了制备 CDs 的荧光发射的随机性，为室温下制备所需发射 CDs 提供了新途径。基于元素稀释效应和氢键的形成，FCDs1 中掺入的少量氟可以缓解聚集诱导的猝灭现象的发生，并使其表现出固态荧光发射。因此，将 FCDs1 作为荧光可视化试剂用于 LFPs 显现。由于 FCDs1 的粒径小，可以实现 LFPs 的高分辨率成像，荧光指纹显示出清晰的线条和明显的细节，甚至可以清楚地观察到汗孔。具有更高氟含量的 FCDs2 荧光发射可红移 70 nm。同时，FCDs2 的荧光光谱与生物分子 CBL 的吸收光谱重叠。因此，基于内滤效应，可以将 FCDs2 作为传感器用于 CBL 的定量检测，检出限为 0.15 μ · mol L^{-1}。

4.2　实验部分

4.2.1　实验仪器

KQ－100B/800KDE 型超声波清洗器（中国昆山市超声仪器有限公司）；

BSA22AS 单盘型分析电子天平（中国北京赛多利斯仪器有限公司）；

TG16－WS 台式高速离心机（中国湘仪实验室仪器开发有限公司）；

2XZ－2 型真空泵（中国临海市谭式真空设备有限公司）；

PB－10 标准型 pH 计（中国北京赛多利斯仪器有限公司）；

Lambda Bio20 紫外可见光分光光度计（美国珀金埃尔默仪器有限公司）；

F－7000 荧光分光光度计（日本日立公司）；

JEM－2100 透射电子显微镜（日本电子株式会社）；

Bruker Dimension icon 原子力显微镜（德国布鲁克公司）；

DZF－6020 型真空干燥箱（中国上海精宏实验设备有限公司）；

One Spectra 红外光谱仪（美国珀金埃尔默仪器有限公司）；

DHG－9037A 电热恒温干燥箱（中国上海精宏实验设备有限公司）；

EscaLab 250Xi X 射线光电子能谱分析仪（美国赛默飞世尔公司）；

YE5A44 型手动可调式移液器（中国上海大龙医疗设备有限公司）；

CHI660E 电化学工作站(中国上海晨华仪器有限公司);

EscaLab 250Xi X 射线光电子能谱分析仪(美国赛默飞世尔公司);

Malvern Nano - ZS 粒度仪(英国马尔文公司);

D8 ADVANCE X 射线衍射仪(德国布鲁克公司);

Horiba XploRA 光谱仪(法国 Jobin Yvon 公司)。

4.2.2 实验试剂

本章所用试剂和品牌如下:

O - PD、4 - 氟 - 1,2 - 苯二胺、对苯醌(p - BQ)和 CBL 购自阿拉丁化学有限公司(中国上海);乙醇、抗坏血酸(AA)、维生素 B_1(VB_1)、维生素 B_6(VB_6)、葡萄糖、丙氨酸(Ala)以及本章使用的其他氨基酸购自国药集团化学试剂有限公司(中国上海)。

除特别声明外,所有试剂皆为分析纯且未经任何前处理。实验用水为二次去离子水(18 MΩ cm)。

4.2.3 UCDs、FCDs1 和 FCDs2 的制备方法

CDs 制备过程如下,将不同质量比的 p - BQ、o - PD 和 4 - 氟 - 1,2 - 苯二胺(0.4 g/1.2 g/0 g, 0.4 g/0.6 g/0.7 g, 0.4 g/0 g/ 1.4 g)加入 120 mL 乙醇中,超声形成均一溶液。在室温下反应 24 小时后,将所得溶液通过 0.22 μm 滤膜以除去大颗粒。通过真空浓缩仪移除溶剂后,将获得的产物分散在水中并离心两次(10 000 $r \cdot min^{-1}$,10 分钟)以洗去不溶成分,并通过冷冻干燥获得粗产物的固体粉末。然后使用二氯甲烷和甲醇的混合物作为洗脱剂将粗产物用硅胶色谱柱法纯化。通过旋转蒸发除去洗脱剂,并在真空干燥箱中进一步干燥。根据组成不同分别命名为 UCDs、FCDs1 和 FCDs2。

4.2.4 UCDs、FCDs1 和 FCDs2 的表征方法

通过循环伏安法测试了 CDs 的电化学性质。其中以玻璃碳电极作为工作电极,铂丝作为辅助电极,Ag/AgCl 作为参比电极。在 0.1 $mol \cdot L^{-1}$ 四正丁基六氟磷酸铵的乙腈溶液中测试了 CDs 样品的循环伏安曲线。CDs 的 HOMO 和 LUMO 能级根据以下公式计算得出:

$$E_{LUMO} = -e(E_{red} + 4.4) \quad (4-1)$$

$$E_{HOMO} = E_{LUMO} - E_g \quad (4-2)$$

E_{LUMO}代表最低未占据分子轨道的能级,E_{HOMO}代表最高占据分子轨道的能级,E_{red}代

表还原电位的起始值，E_g 代表带隙。

此外，本章通过 U－3900 紫外可见分光光度计使用 1 cm 光程的比色皿，在扫描间隔为 1 nm 条件下测得紫外－可见（UV－vis）吸收光谱。荧光光谱是通过 F－7000 荧光光谱仪使用 1 cm 光程的比色皿测得，激发和发射狭缝均设置为 10 nm。傅里叶红外（FT－IR）光谱使用 Nicolet－6700 红外光谱仪采用溴化钾压片法测定并记录 1 000 ~ 4 000 cm^{-1} 的数据。X 射线光电子能谱（XPS）在配备 Al Kα 280.00 eV 激发光源的 ES-CALAB 250 表面分析系统上进行测得。透射电子显微镜（TEM）图像通过 JEM－2100 高分辨透射电子显微镜测得，加速电压为 200 kV。

4.2.5　UCDs、FCDs1 和 FCDs2 的稳定性测试

光稳定性测试：在 365 nm 波长下连续照射 UCDs、FCDs1 和 FCDs2 溶液 3 600 s，并记录三种 CDs 样品在最大荧光发射时的强度值。每组样品测试均重复 3 次。

储存稳定性测试：在室温条件下将 UCDs、FCDs1 和 FCDs2 溶液保存 7 天，并每 24 小时在最佳激发波长下测试 CDs 样品在最大荧光发射时的强度值。每组样品测试均重复 3 次。

4.2.6　FCDs1 用于潜指纹显现

LFPs 样品来自不同性别的志愿者。操作过程如下：用肥皂将手清洗干净，在空气中晾干，然后在鼻子或额头上轻轻擦拭手指使油脂留在手指上。将手指压在不同的基材上形成指纹，之后把 FCDs1 粉末撒在指纹样本上并吹去多余的粉末。最后，在紫外灯下观察 LFPs。

4.2.7　FCDs2 用于钴胺素定量检测

通过以下步骤用 FCDs2 作为荧光探针进行 CBL 测定。将不同浓度的 CBL 溶液与 3 mL FCDs2 溶液混合（12.5 $\mu g \cdot mL^{-1}$，在 pH 值为 7.4 的磷酸盐缓冲盐水（PBS）缓冲液中制备）并定容至 4 mL。随后，使用 540 nm 的激发波长测试荧光发射光谱。以 CBL 浓度为横坐标，荧光猝灭率 $(F_0-F)/F_0$ 为纵坐标绘图，其中 F_0 和 F 分别代表添加 CBL 前后 FCDs2 的荧光强度。此外，在相同条件下将 CBL 替换为干扰物研究了 FCDs2 的选择性。所有测试均重复 3 次。

为了考察 FCDs2 在实际样品中检测 CBL 的可行性，本研究以 FCDs2 作为探针通过加标回收法对市售功能饮料和药片中的 CBL 进行了定量分析。功能饮料购买于本地超

市,开瓶稀释后用于 CBL 检测。CBL 药片购买于本地药店,将 CBL 药片捣碎,溶解在 10 mL生理盐水中,然后以5 000 r · min^{-1}转速离心 20 分钟除去不溶物。将上清液收集并稀释用于 CBL 检测。

4.3 结果与讨论

4.3.1 UCDs、FCDs1 和 FCDs2 的制备与表征

氟掺杂的表面态调控法是指:通过调控不同的氟含量而获得不同表面态 CDs 的一种方法。在本研究中,CDs 的形成过程如图 4 – 1 所示,在控制 p – BQ 用量不变的情况下,通过改变 o – PD 及其衍生物 4 – 氟 – 1,2 – 苯二胺的用量制备出了三种不同氟含量的 o – PD 衍生 CDs。事实上,o – PD 及其衍生物 4 – 氟 – 1,2 – 苯二胺的 – NH_2 基团均具有很高的活性,因此在空气中可以通过氧化/聚合反应生成各种氧化产物。同时,在 p – BQ 存在的情况下,o – PD/4 – 氟 – 1,2 – 苯二胺及它们氧化产物上的 – NH_2 能与p – BQ 的 C = O 之间发生席夫碱反应形成 C = N。同时,席夫碱反应释放的热量进一步促进了前驱体的聚合和氧化反应,从而促进了 CDs 的形成[5]。

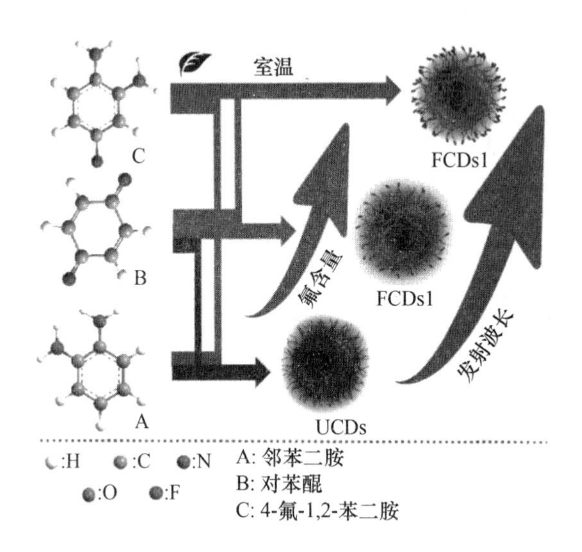

图 4 – 1 以 o – PD/4 – 氟 – 1,2 – 苯二胺和 p – BQ 为前驱体在室温下制备 CDs 的示意图

TEM 是把经加速和聚集的电子束投射到非常薄的样品上,电子与样品中的原子碰撞而改变方向,从而产生立体角散射。散射角的大小与样品的密度、厚度相关,因此可以形成明暗不同的影像,影像将在放大、聚焦后在成像器件上显示出来。TEM 可以看到

在光学显微镜下无法看清的小于 0.2 μm 的细微结构,这些结构称为亚显微结构或超微结构。因此,TEM 被认为是确定 CDs 形状、形态和大小的有效手段。图 4－2(a),图 4－2(b)和图 4－2(c)中的 TEM 图像显示三种 CDs 呈单分散状态,平均尺寸分别为 5.4 nm、5.3 nm 和 5.2 nm。它们都是圆球形的纳米点,分布均匀且无团聚。

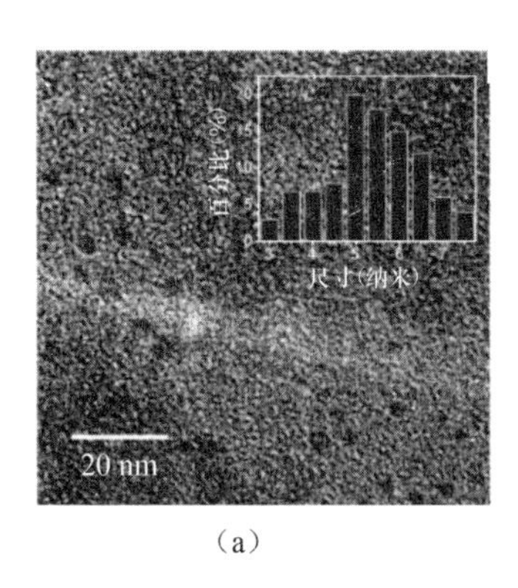

(a)

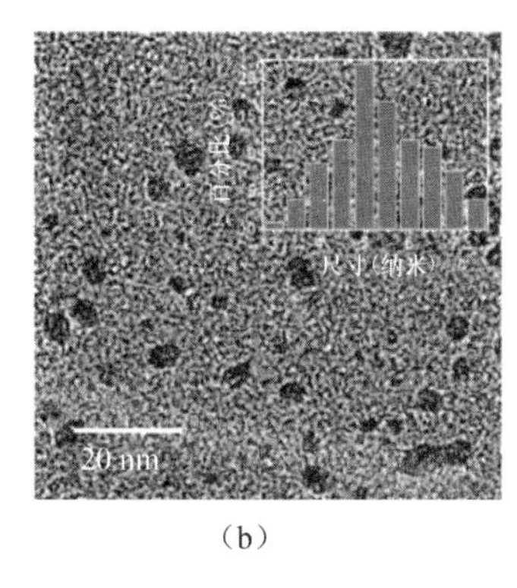

(b)

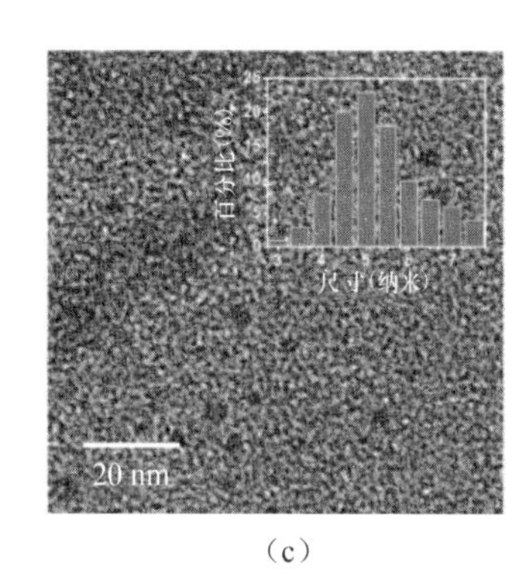

(c)

图 4－2　UCDs、FCDs1 和 FCDs2 的 TEM 图像

组成分子的化学键或官能团上的原子处于不断振动的状态,其振动频率与红外光的振动频率相当。当用红外光照射分子时,分子中的化学键或官能团可发生振动吸收,不同的化学键或官能团吸收频率不同,在 FT－IR 上将处于不同位置,从而可获得分子中含有何种化学键或官能团的信息。通过 FT－IR 光谱分析表征 CDs 上的官能团。从图 4－3(a)中可以看出,三种 CDs 的 FT－IR 谱图显示了特征吸收带,对应于 N－H/O－H 在 3 500～3 150 cm^{-1} 处的伸缩振动,C－H 在 2 915 cm^{-1} 处和 2 833 cm^{-1} 的伸缩振动,在 1 622 cm^{-1} 处 C＝N 的伸缩振动,在 1 490 cm^{-1} 处 C＝C 的伸缩振动,在 1 239 cm^{-1} 处 N－H 的弯曲振动以及在 1 109 cm^{-1} 处 C－O 的伸缩振动[27-30]。此外,FCDs1 和 FCDs2 在 1 158 cm^{-1} 和 1 293 cm^{-1} 处出现两个吸收峰,而 UCDs 则未观察到。该峰归因于 C－F 伸缩振动,这表明氟成分已成功掺杂到 CDs 结构中[31]。与 FCDs1 相比,在 FT－IR 光谱中 FCDs2 的 C－F 的强度更明显,这表明 FCDs2 上存在更多的 C－F 基团。此外,FT－IR 光谱证实了三种 CDs 样品中存在席夫碱结构(C＝N 谱带),进一步揭示了 CDs 的形成中存在席夫碱反应。通过 XPS 光谱分析了三种 CDs 样品的组成和表面态。在 XPS 中[图 4－3(b)],三种 CDs 在 284 eV、399 eV 和 532 eV 存在 3 个峰,分别对应于 C 1s、N 1s 和 O 1s[32-34]。除了 C、N 和 O 外,在 FCDs1 和 FCDs2 中还观察到 F(685 eV)的存在[35]。同时,FCDs1 和 FCDs2 中氟的含量分别为 1.84% 和 2.40%。

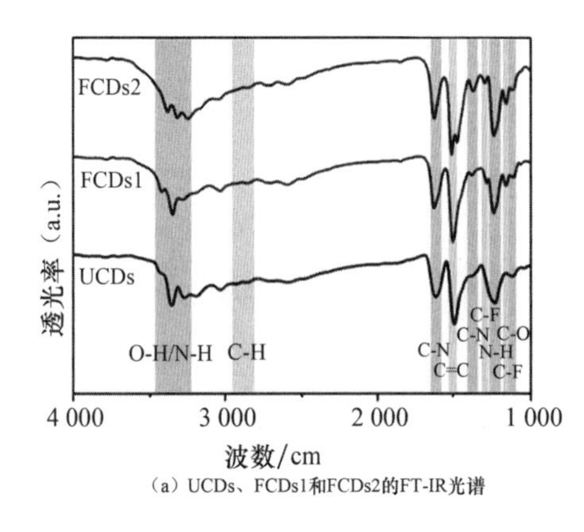

（a）UCDs、FCDs1和FCDs2的FT-IR光谱

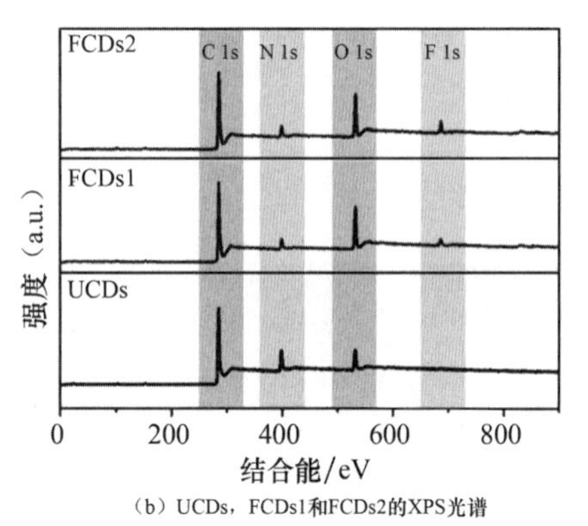

（b）UCDs，FCDs1和FCDs2的XPS光谱

图 4-3

通过高分辨率 XPS 进一步分析三种 CDs 样品的化学成分。UCDs 的 C 1s 光谱图 4-4(a)可以分成 4 个结合能峰,分别对应于 C=C/C-C、C-N、C-O 和 C=N/C=O。图 4-4(b)和图 4-4(c)分别为 FCDs1 和 FCDs2 的 C 1s XPS 光谱。对于这两种 CDs,C 1s光谱在 284.5、285.2、286.3、287.3 和 288.5 eV 处显示 5 个明显的峰,分别对应于 C=C/C-C、C-N、C-O、C-F 和 C=N/C=O[36-38]。在 F 1s 光谱图 4-4(d)和图 4-4(e)显示了 FCDs1 和 FCDs2 中存在两种 F 键类型分别为:半离子 C-F 键和共价 C-F 键[39,40]。三种 CDs 的 N 1s 谱和 O 1s 谱展现了相同的结合能峰。如图 4-5(a),图 4-5(b)和图 4-5(c)所示,N 1s 可以分为 2 个峰,分别对应于 C-N=C 和 $-NH_2$[13]。而三种 CDs 样品的 O 1s 光谱图 4-5(d),图 4-5(e)和图 4-5(f)分别包含 2 个峰,对应于 C=O 和 C-O[41]。

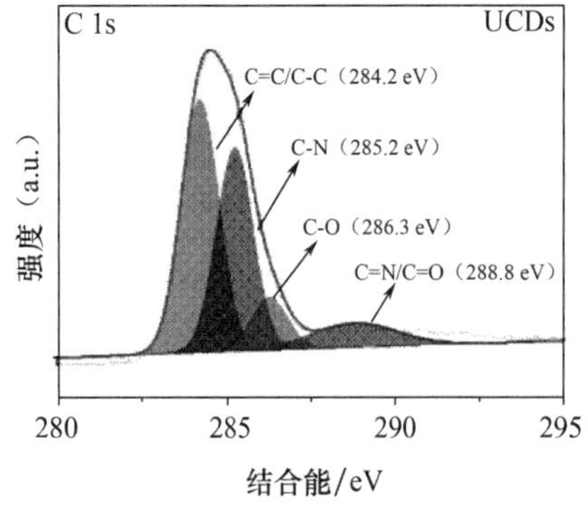

（a）UCDs，FCDs1和FCDs2的高分辨率C 1s XPS

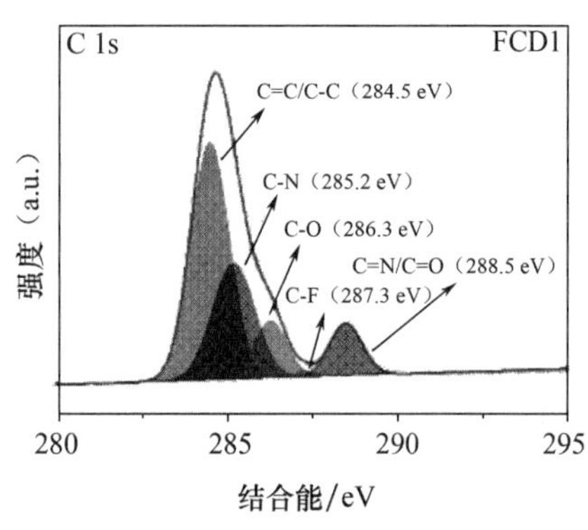

（b）UCDs，FCDs1和FCDs2的高分辨率C 1s XPS

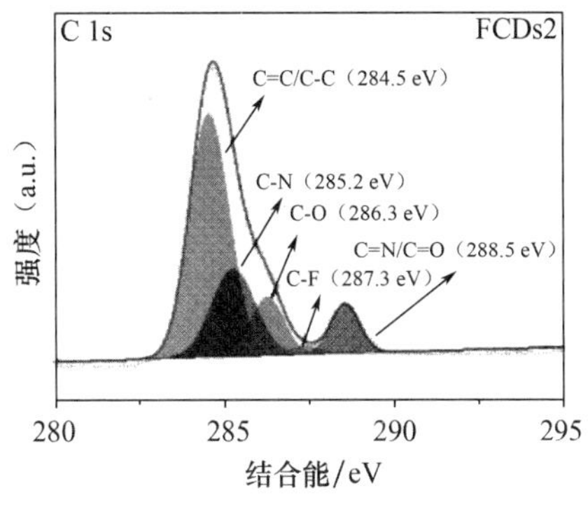

（c）UCDs，FCDs1和FCDs2的高分辨率C 1s XPS

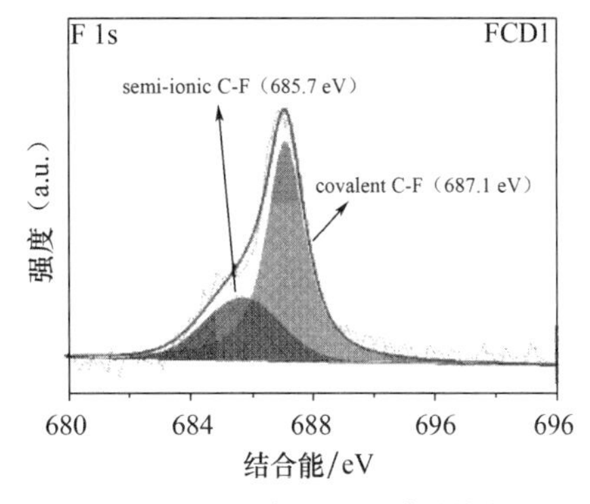

（d）UCDs，FCDs1和FCDs2的高分辨率F 1s XPS

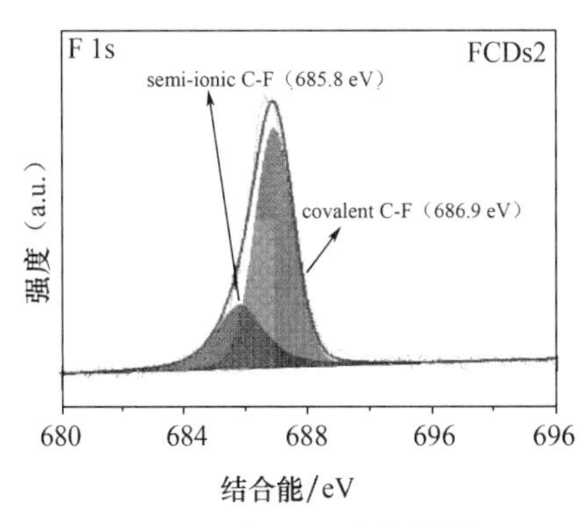

（e）UCDs，FCDs1和FCDs2的高分辨率F 1s XPS

图 4 – 4

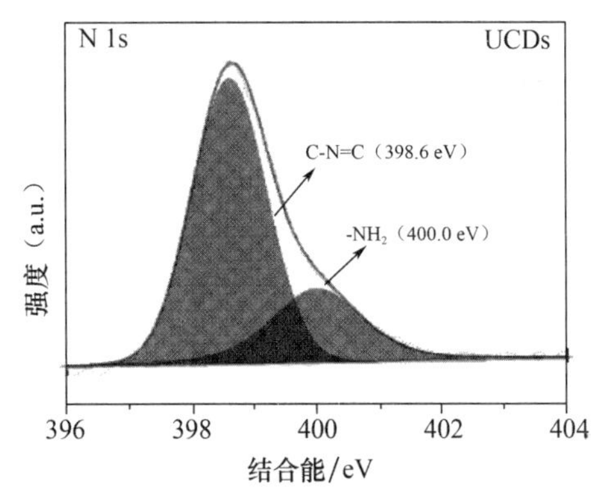

（a）UCDs，FCDs1和FCDs2的高分辨率N 1s XPS

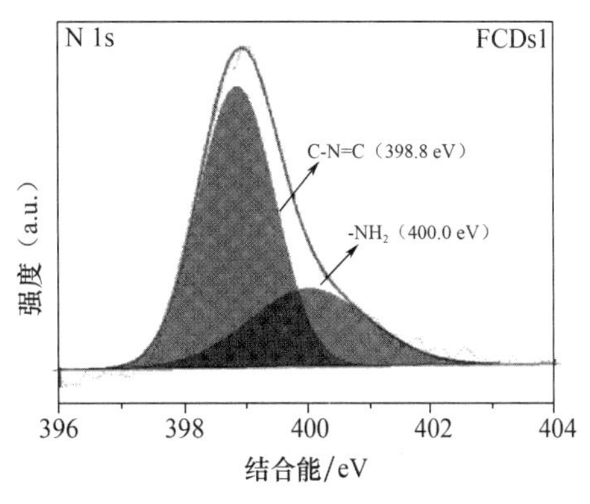

（b）UCDs，FCDs1和FCDs2的高分辨率N 1s XPS

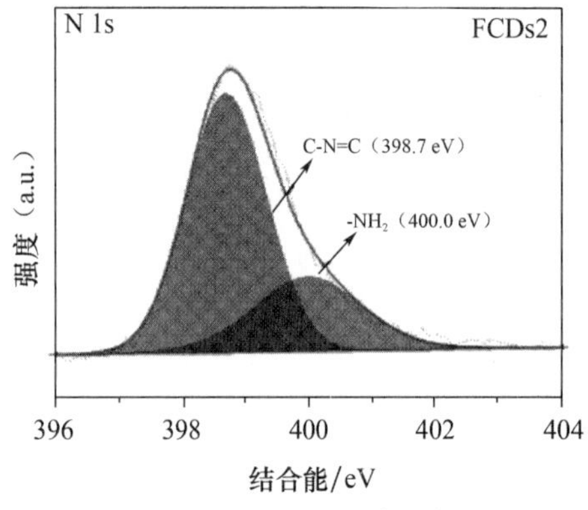

（c）UCDs，FCDs1和FCDs2的高分辨率N 1s XPS

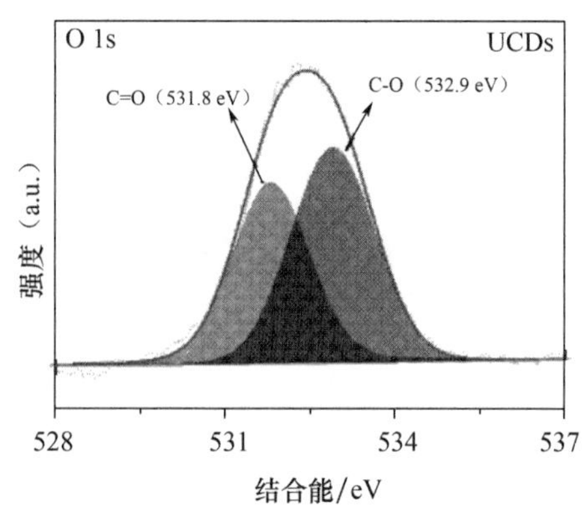

（d）UCDs，FCDs1和FCDs2的高分辨率O 1s XPS

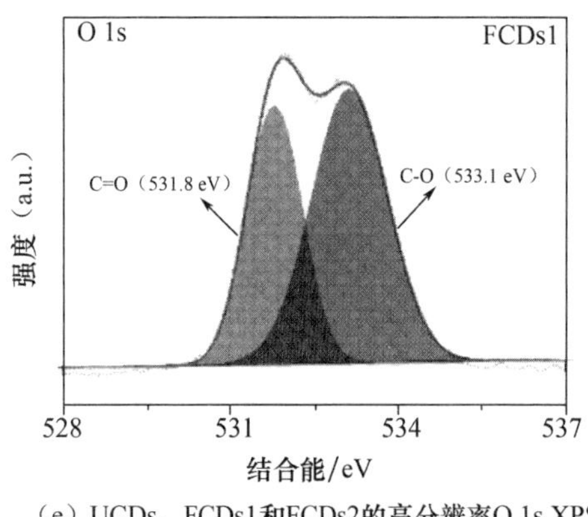

（e）UCDs，FCDs1和FCDs2的高分辨率O 1s XPS

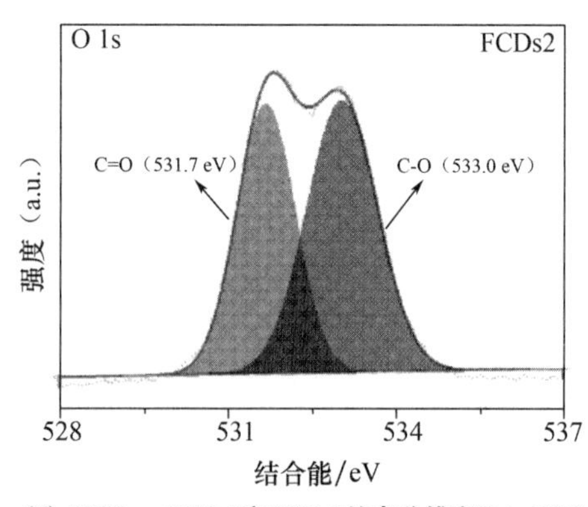

（f）UCDs，FCDs1和FCDs2的高分辨率O 1s XPS

图 4 -5

4.3.2 UCDs、FCDs1 和 FCDs2 的光学性质

图 4 -6 展示了 3 种 CDs 的光学表征结果。在有机化合物分子中有形成单键的 σ 电子、有形成双键的 π 电子、有未成键的孤对 n 电子。当分子吸收一定能量的辐射能时，这些电子就会跃迁到较高的能级，此时电子所占的轨道称为反键轨道，而这种电子跃迁同内部的结构有密切的关系，这就产生了 UV - Vis 光谱。在 UV - vis 吸收光谱中，3 种 CDs 在 200 ~ 360 nm 高能区中均表现出了 C = C 引起的 $\pi-\pi^*$ 跃迁和 C = O/C = N 引起的 $n-\pi^*$ 跃迁的吸收带。除了这些吸收带，在 360 ~ 700 nm 的低能区也观察到了吸收带，这归因于共轭 C = N 的 $n-\pi^*$ 跃迁[42,43]。对于 CDs 来讲，无论是基础研究还是实际应用，它的荧光行为都是最令人关注的特征。控制变量法用于系统地研究氟含量对 CDs 荧光特征的影响。将 p - BQ 的质量固定为 0.4 g，通过改变 o - PD 和 4 - 氟 - 1，

2-苯二胺的质量比调节氟含量。如图4-6(a)所示,无氟成分的UCDs作为空白组在530 nm出观察到了最大发射峰。随着氟成分的增加,如图4-6(b)和4.6(c)所示,FCDs1和FCDs2的最大荧光发射峰分别红移至555 nm和600 nm。此外,在不同激发波长下,UCDs显示荧光行为的激发波长依赖性。而与之相反,当施加不同的激发波长时,FCDs1和FCDs2的最大荧光发射位置不会发生变化。事实上,CDs的这种荧光行为通常与其表面态有关,在这里可能是由氟掺杂引起的CDs表面态变化所引起的[36]。特别地,FCDs1粉末表现出固态荧光发射特性(图4-7),而UCDs和FCDs2在固态时却未观察到此现象。由此可以看出,氟原子的量在改善CDs的固态荧光中也起到了重要作用。

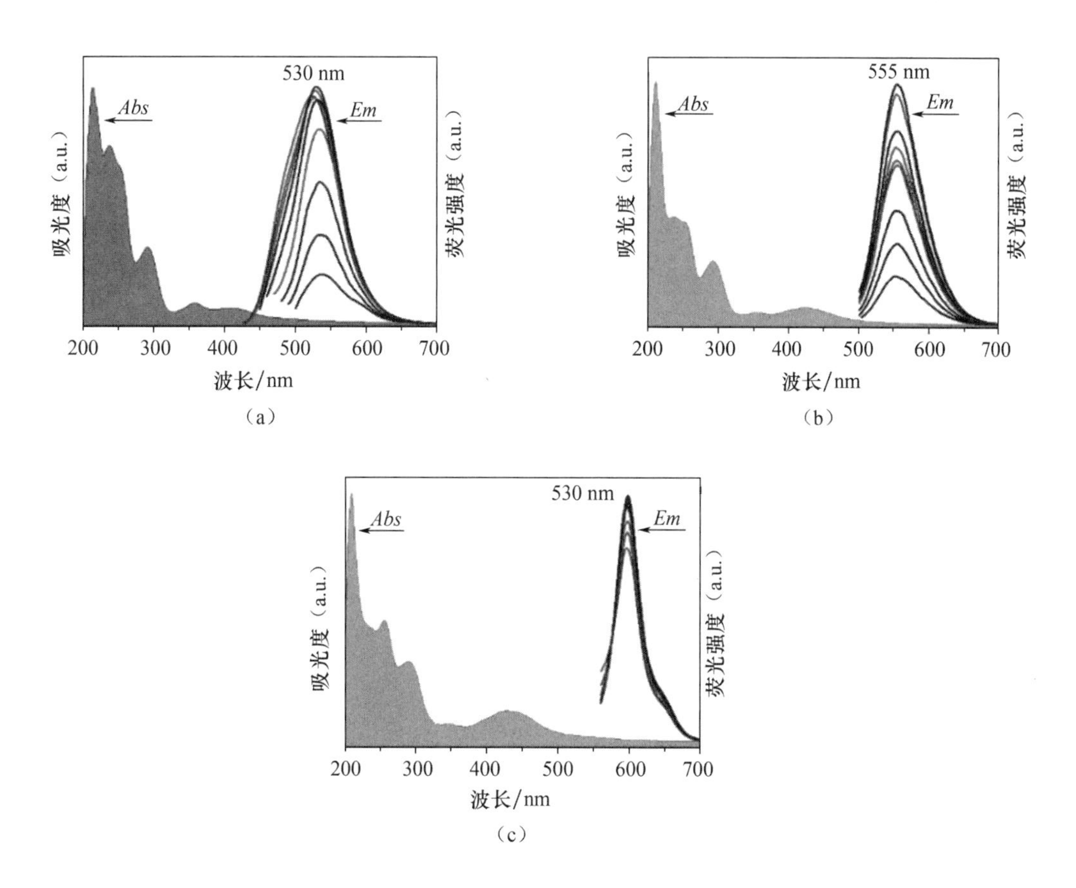

图4-6 UCDs,FCDs1和FCDs2的UV-vis吸收光谱和不同激发波长荧光发射光谱

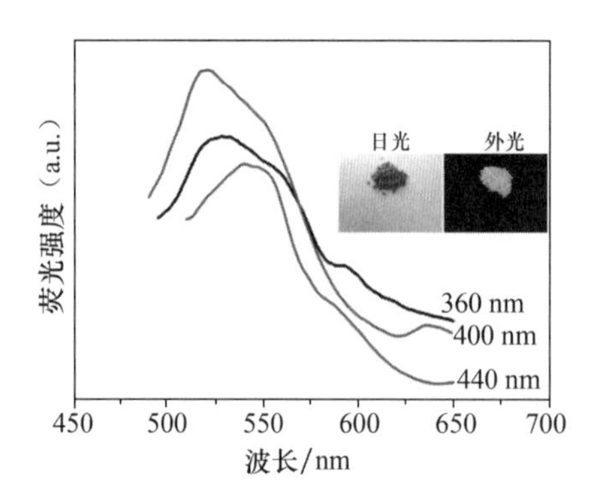

图 4-7　在不同激发波长下 FCDs1 的固态荧光发射光谱

测试了 CDs 的荧光寿命,3 种 CDs 的荧光衰减曲线图 4-8(a)可以用单指数函数拟合,拟合后计算 UCDs、FCDs1 和 FCDs2 的荧光寿命分别为 4.73 ns、2.22 ns 和 2.04 ns。单指数荧光寿命表明样品包含单重态荧光中心[44],不同的荧光寿命进一步 CDs 证实了不同类型的电子激发态[45]。为了更直观的理解 3 种 CDs 的荧光差异,本研究中计算了 CDs 的色坐标,图 4-8(b)计算后 UCD、FCDs1 和 FCDs2 的 CIE 色坐标分别为(0.262 4,0.567 4),(0.290 7,0.615 5)和(0.475 7,0.518 3),这进一步揭示了 3 种 CDs 的不同发射行为。CDs 的荧光稳定性在其实际应用中起着关键作用,本工作通过氙气灯连续照射 3 600 s 后观察到 3 种 CDs 的荧光强度基本不变,说明它们具有极好的光稳定性[图 4-9(a),图 4-9(b)和图 4-9(c)]。此外,o-PD 衍生 CDs 在室温下存放 7 天后仍显示出稳定的荧光发射图 4-9(d),图 4-9(e)和图 4-9(f)。

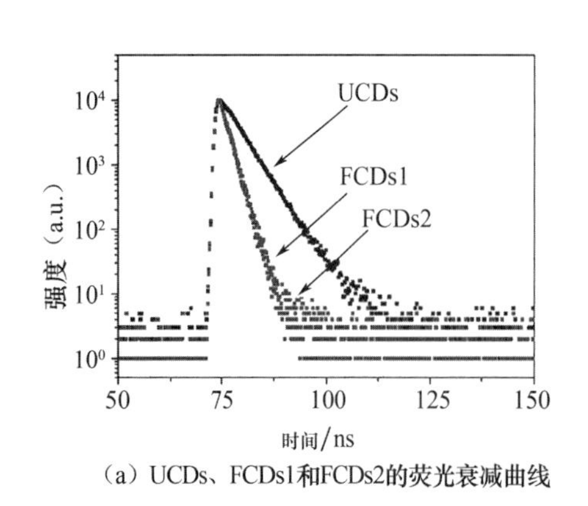

(a) UCDs、FCDs1和FCDs2的荧光衰减曲线

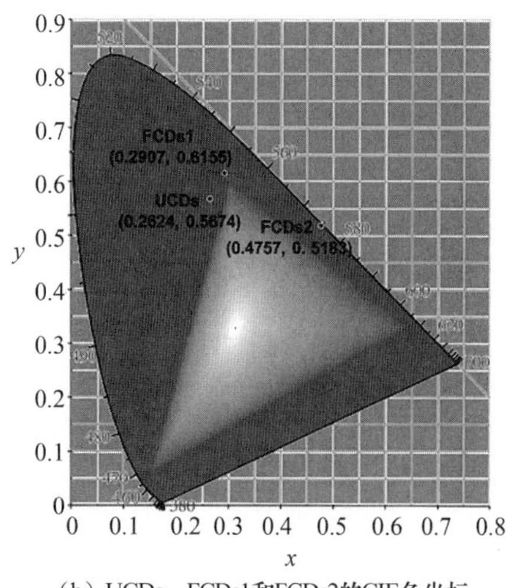

(b) UCDs、FCDs1和FCDs2的CIE色坐标

图 4-8

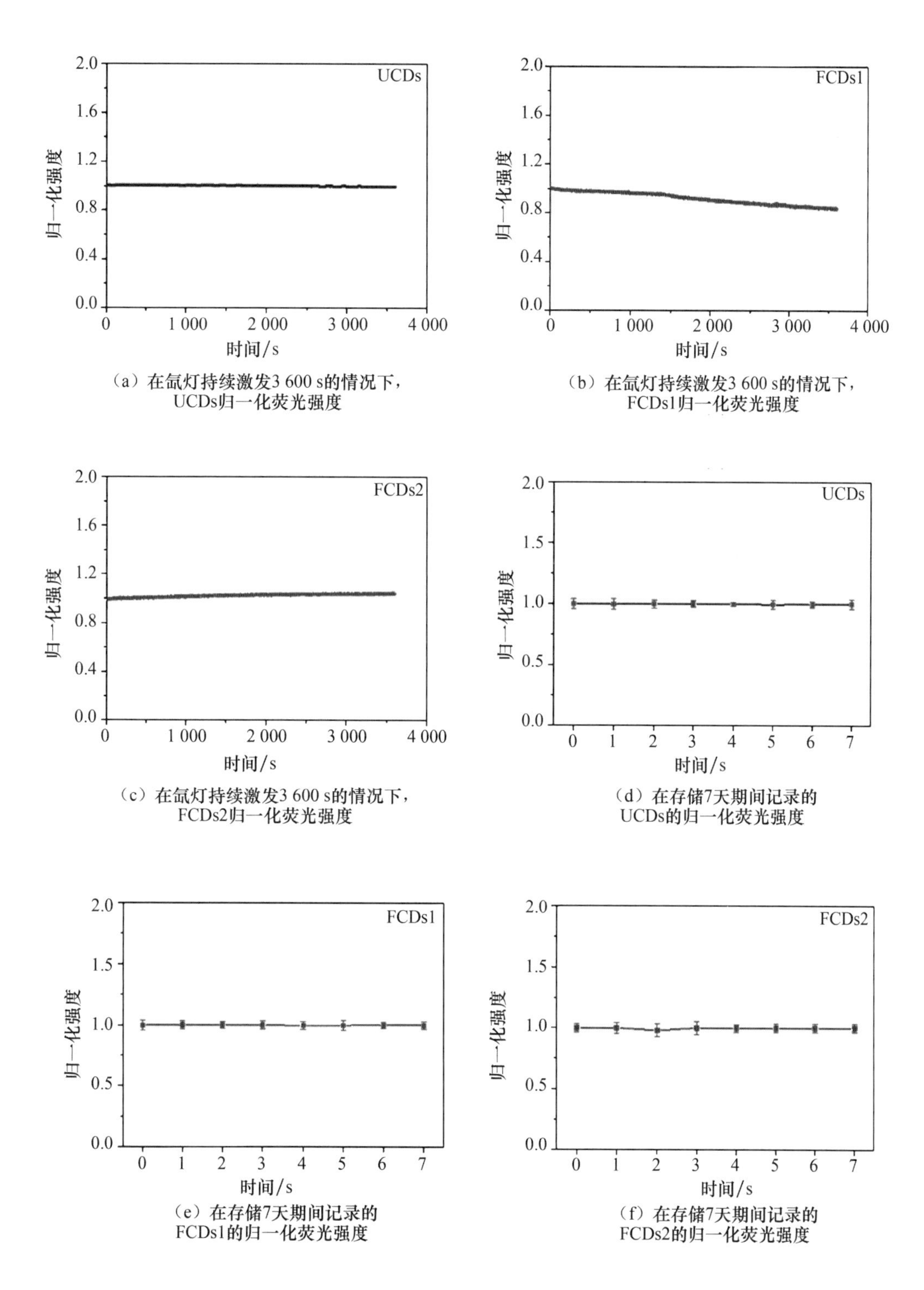

（a）在氙灯持续激发3 600 s的情况下，UCDs归一化荧光强度

（b）在氙灯持续激发3 600 s的情况下，FCDs1归一化荧光强度

（c）在氙灯持续激发3 600 s的情况下，FCDs2归一化荧光强度

（d）在存储7天期间记录的UCDs的归一化荧光强度

（e）在存储7天期间记录的FCDs1的归一化荧光强度

（f）在存储7天期间记录的FCDs2的归一化荧光强度

图 4－9

4.3.3 UCDs、FCDs1 和 FCDs2 的荧光发射机理

为了研究 CDs 的荧光发射机理,本工作通过循环伏安法测试 3 种 CDs 的电化学性质,并计算了 3 种 CDs 的 HOMO 和 LUMO 能级。由图 4 - 10(a)、图 4 - 10(b)和图 4 - 10(c)可知,UCDs、FCDs1 和 FCDs2 的 E_{red}分别为:-0.56 V、-0.54 和 -0.50 V。因此,通过公式 4 - 1 计算 UCDs、FCDs1 和 FCDs2 的 E_{LUMO}分别为:-3.84 eV、-3.86 eV 和 -3.90 eV。通过 3 种 CDs 紫外漫反射光谱得出 E_g 分别为:3.49 eV、3.21 和 2.98 eV [图 4 - 10(d)、图 4 - 10(e)和图 4 - 10(f)]。因此,计算得出 E_{HOMO} 为 -7.33 eV、-7.07 eV 和 -6.88 eV。可以看出,UCDs、FCDs1 和 FCDs2 显示不同的 E_{LUMO}和 E_{HOMO}。从图 4 - 11 可以看出,当 CDs 的荧光发射红移时,相应的 E_g 随 HOMO 水平的提高和 LUMO 水平的降低而降低。

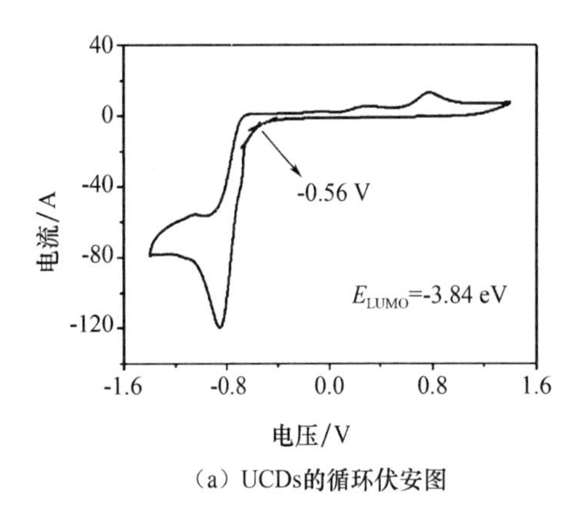

(a) UCDs的循环伏安图

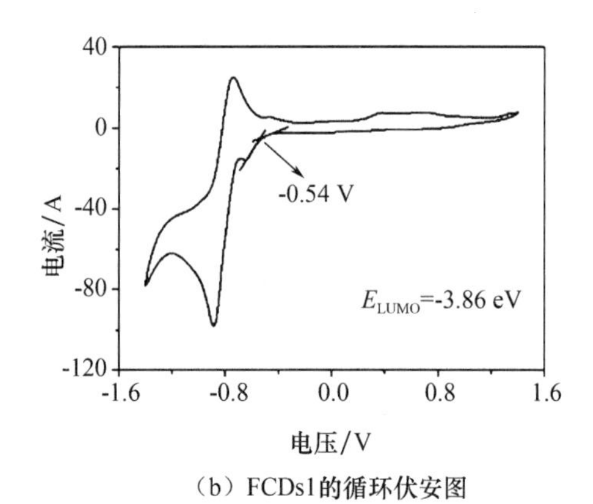

(b) FCDs1的循环伏安图

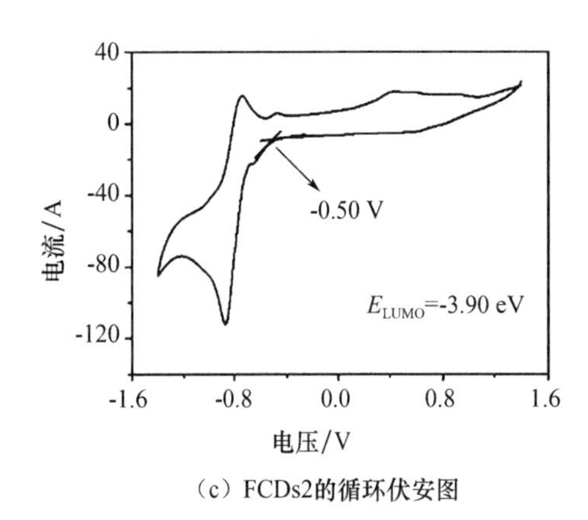

(c) FCDs2的循环伏安图

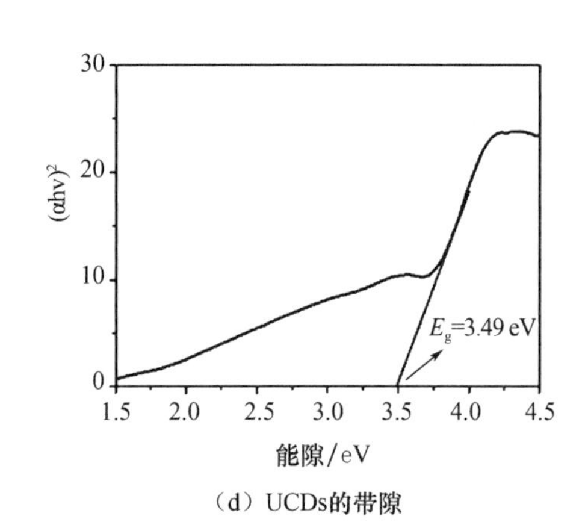

(d) UCDs的带隙

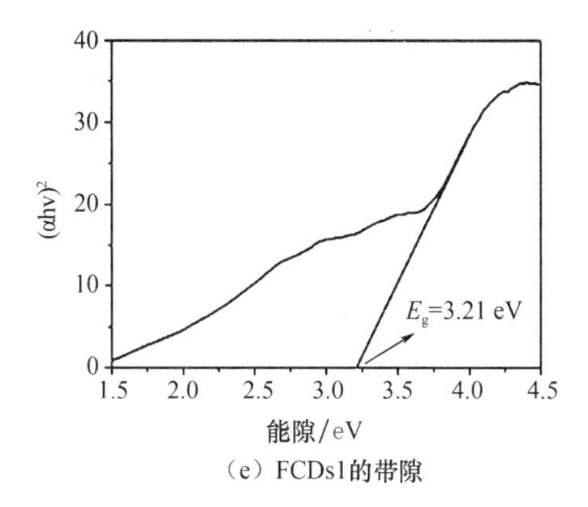

（e）FCDs1的带隙

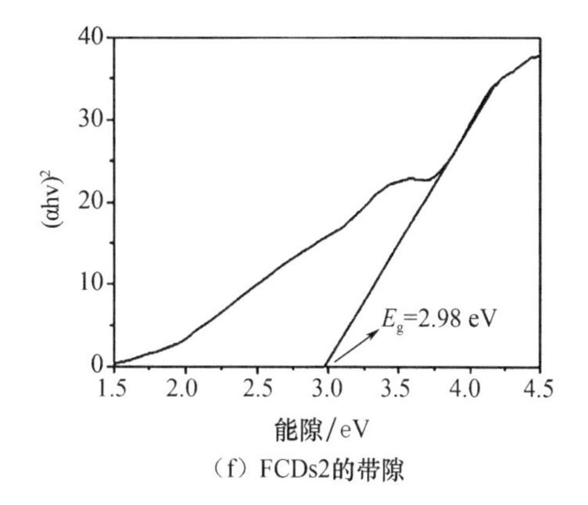

（f）FCDs2的带隙

图 4－10

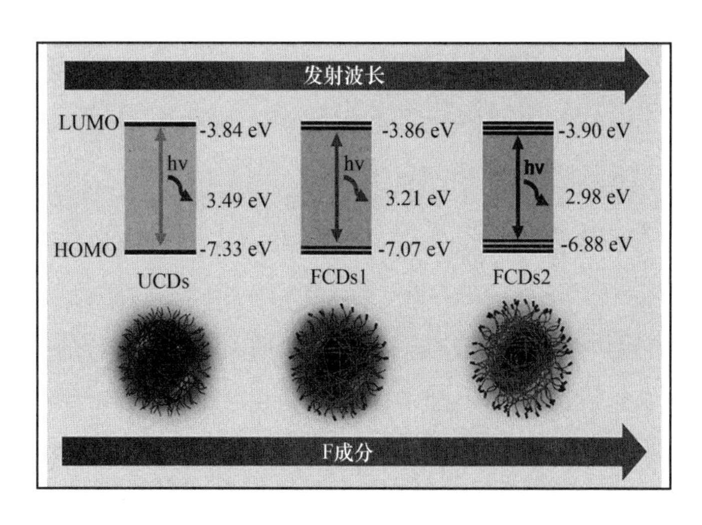

图 4－11　UCDs、FCDs1 和 FCDs2 的能级图

CDs 的结构是如何引起能级变化并影响荧光发射的是十分值得思考的。氟是常见的元素之一，具有最大的电负性，可以强烈吸引相邻电子并增加负电荷和正电荷的分离程度。因此，在氟的参与下荧光材料的激发/发射波长会发生极大改变[35]。从上述表征可以看出，制备的 3 种 o－PD 衍生 CDs 的荧光发射与氟成分密切相关，3 种 CDs 荧光发射随着氟含量的增加红移。事实上，这是由于 CDs 上的氟原子能够增加 HOMO 上的电子密度而引起的荧光发射发生红移[46]。因此，得出结论，这些室温 CDs 的荧光发射与氟的含量密切相关，即 CDs 中的氟元素会调节 HOMO 和 LUMO 的能级，从而导致 CDs 的荧光发生红移。此外，CDs 的氟含量与它们的红移荧光发射正相关，也就是说 CDs 中较高氟含量有助于带隙变窄，导致这些 CDs 的荧光发射红移。

在本研究中，当 F 掺杂量仅为 2.40% 时，CDs 的发射波长可以红移达 70 nm，这证明了该方法对 CDs 的荧光发射红移调控的高效性[35]。如图 4－12 所示，当将 FCDs2 的制备温度提高到 80 ℃时，CDs 的荧光发射峰蓝移到 556 nm。随着温度升高到 140 ℃，荧光发射峰的继续蓝移，最强的荧光发射在 545 nm。这足以表明温度可以影响基于氟掺杂的表面态控制法对 CDs 荧光红移效率，这一发现将为未来设计与 CDs 荧光发射调控有关的方法提供新的思路。

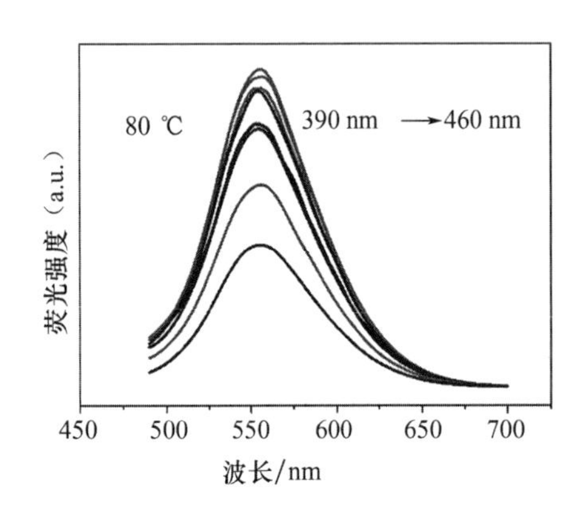

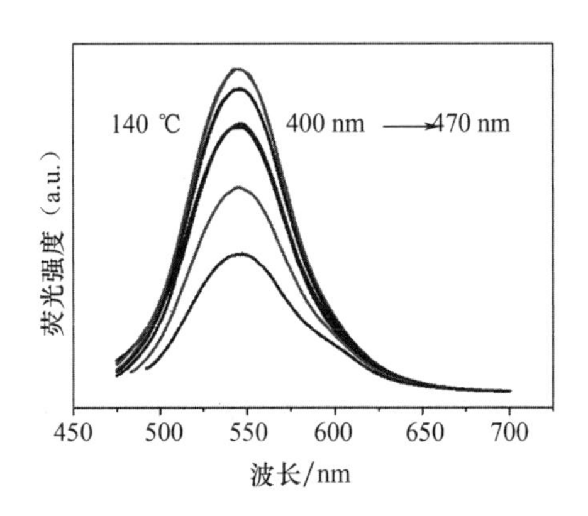

图 4－12　在 80 ℃和 140 ℃下制备的 FCDs2 不同激发波长的荧光发射光谱

由于聚集诱导荧光淬灭作用的存在，大多数 CDs 在固态时都会失去荧光。但是对于本章制备的 CDs 来讲，固态荧光与氟含量有很大的关系，这表明它们的固态荧光是一种特殊的机理，本工作认为 3 种 CDs 固态荧光的变化可能是由以下原因引起的：作为无氟成分的 UCDs，它与大多数 CDs 相同，由于荧光共振能量转移或直接 π－π 相互作用的影响，其在固态时会受到严重的聚集诱导淬灭而无荧光发射[47]。当 FCDs1 的结构中存在氟成分时，CDs 的荧光特性将会受到氟原子的调节，这些基于少量氟原子的荧光发射中心会被稀释而无聚集诱导淬灭的发生[48]。此外，在 FCDs1 上存在大量的 $-NH_2$ 和 C－F 基团可以促进氢键的形成[31]。氢键可以使相邻 CDs 之间保持一定距离，避免它们之间的直接接触和过多的荧光共振能量转移的发生，从而在固态下抑制了荧光淬灭[49]。因此，在固态时它们的荧光发射不会相互干扰。然而，随着氟含量的增加，FCDs2 将形成更多的 C－F 基团，而这些基团会与自身 CDs 的 $-NH_2$ 基团结合形成氢键，这将减少 CDs 之间的氢键数量并缩短 CDs 之间的距离。最后，受到浓度猝灭作用的影响 FCDs2 的固态荧光消失。

4.3.4　基于 FCDs1 粉末的潜指纹显现

考虑到 FCDs1 独特的固态荧光发射性、低毒性、小粒径和丰富的官能团，本工作探索了 FCDs1 用于潜指纹可视化应用的可行性。图 4 - 13(a)所示,在 365 nm 照射下男性志愿者在纸上的指纹经 FCDs1 处理后的荧光图像,所得荧光指纹的线条清晰,细节明显。在放大的图像中,核心、终端、点、分叉、三角区、孤立区和疤痕可以被观察到。除此之外,还可以观察到汗孔。实际上,即使指纹不完整或被损坏,汗孔也能用于个人的识别[50]。通常,指纹的唯一性由局部脊特征或其关系确定。然而,男性和女性的脊的大小不仅有区别,而且孔的大小也从 88 μm 到 220 μm 不等[19]。因此,本工作选择了在相同基质上保留的女性志愿者指纹用于观察。图 4 - 13(b)所示为女性志愿者在纸上留下指纹的荧光图像显影,从图中可以清晰地看到脊线流动、脊取向和脊图案。放大的区域图像展示了指纹细节,包括:核心、终端、点、分叉、三角区、孤立区、湖状、疤痕和汗孔。这些详细的结构对于基于指纹的个人识别至关重要。另外,在锡箔和陶瓷上也进行了相同志愿者的 LFPs 采集。如图 4 - 13(c)和图 4 - 13(d)所示,对应与图 4 - 13(a)和图 4 - 13(b)的相同位置可以观察到同样的特征曲线,这表明 FCDs1 粉末可应用于不同基材上 LFPs 的可视化。此外,从犯罪现场收集指纹时,LFPs 可视化的稳定性也非常重要。如图 4 - 14 所示,在存储一段时间后的 LPFs 荧光图像仍然清晰明亮且可以观察到极高分辨率的荧光图像,这表明 FCDs1 具有很高的稳定性和实用性。

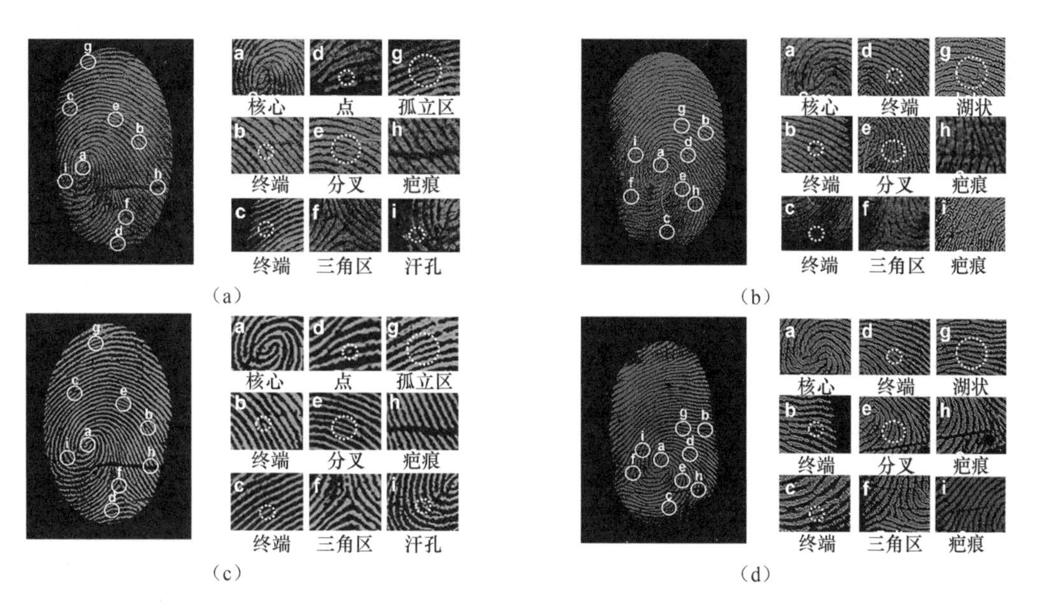

图 4 - 13　在 365 nm 下的 FCDs1 染色的 LFPs 的荧光图像显影

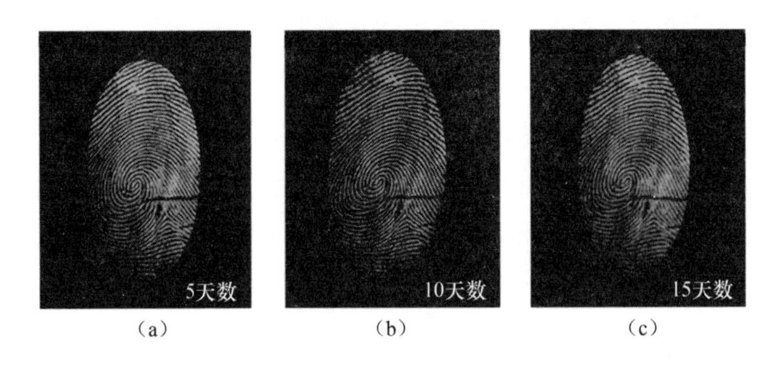

图 4－14　不同时间存储后的 LFPs 的荧光图像

4.3.5　基于 FCDs2 的钴胺素荧光检测

CBL 作为含钴生物分子，不仅对神经细胞的修复发挥作用，而且对红细胞的生长也有重要贡献。缺乏 CBL 会导致严重的贫血等疾病，而过量摄取 CBL 也会引起负面影响。研究发现，CBL 可以有效地抑制 FCDs2 的荧光发射。因此，本工作以 FCDs2 为探针用于荧光检测 CBL。图 4－15（a）展示了添加一系列不同浓度 CBL 后（0～150 μ·mol L^{-1}），FCDs2 的荧光强度变化。从图 4－15（b）可以看出，FCDs2 的荧光随着 CBL 浓度的增加而降低。当 CBL 浓度在 0.5～60 μ·mol L^{-1} 的范围内，淬灭率 $(F_0-F)/F_0$ 与 CBL 浓度之间具有良好的线性关系，线性方程为：$(F_0-F)/F_0=0.0076C+0.02$，($R^2=0.994$)。当信噪比为 3 时，检出限为 0.15 μ·mol L^{-1}。与其他 CBL 检测方法相比（表 4－1），该传感体系具有操作简单、线性范围宽、灵敏度高等优点。

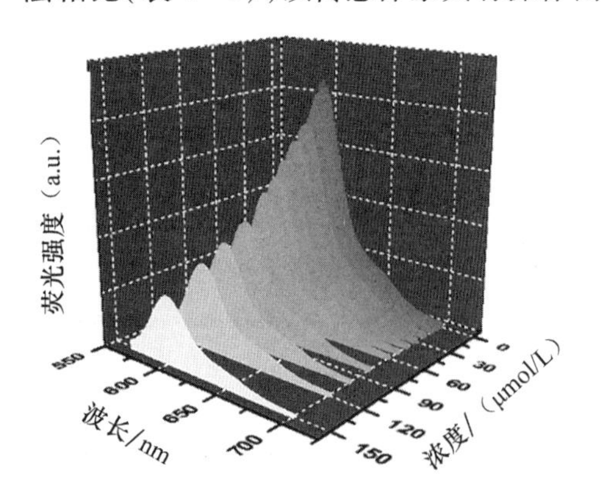

（a）加入不同浓度的CBL后的FCDs2的荧光发射光谱

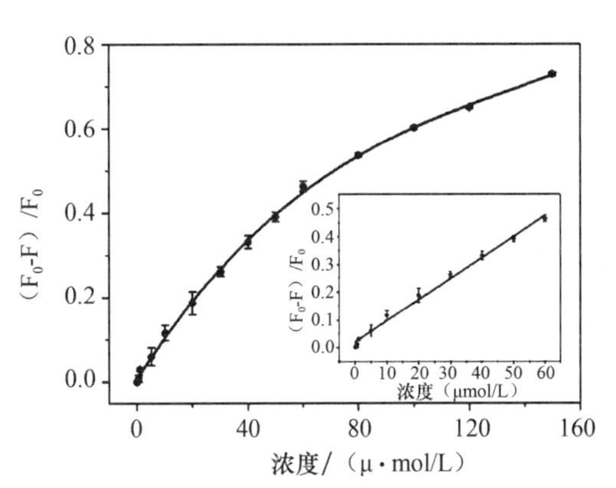

（b）荧光淬灭率 $(F_0-F)/F_0$ 与CBL浓度之间的关系［插图为 $(F_0-F)/F_0$ 与CBL浓度之间线性关系］

图 4－15

表 4 – 1　检测 CBL 分析方法的比较

分析方法	线性范围/(μ · mol L^{-1})	检出限/(μ · mol L^{-1})	参考文献
荧光法	—	0.32	[51]
荧光法	0.75 – 100	0.2	[52]
荧光法	1 – 65,70 – 140	0.62	[53]
荧光法	0.5 – 16	0.158	[54]
高效液相色谱法	1.84 – 9.22	0.92	[55]
电化学法	10 – 60	5	[56]
荧光法	0.5 – 60	0.15	本工作

随后,考察了 FCDs2 传感器对 CBL 的选择性。如图 4 – 16 所示,在相同条件下几种重要的生物分子和电解质对 FCDs2 荧光的影响可忽略不计。这些观察结果表明 FCDs2 不仅对 CBL 灵敏性,而且对检测 CBL 有选择性。此外,本工作对 FCDs2 在实际样品中检测 CBL 的实用性进行了研究。选择功能饮料和 CBL 片剂作为实际样品,测试结果汇总在表 4 – 2 中。从表中可以看出,使用加标回收法验证了方法的准确性,CBL 的回收率在 95.0% 至 99.9% 之间,相对标准偏差为 2.1% ~7.2% 。以上结果证实 FCDs2 适用于实际样品中 CBL 定量分析。

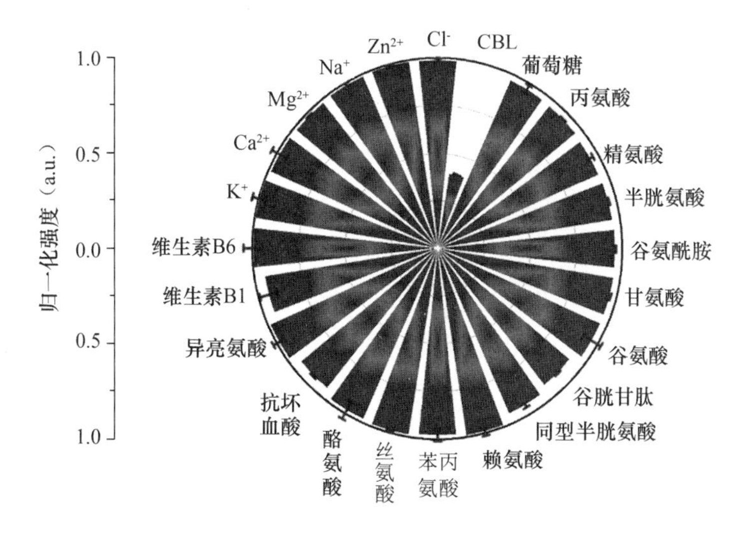

图 4 – 16　FCDs2 对几种重要生物分子和电解质的荧光响应性

表 4 -2　实际样品中 CBL 的测定结果

样品名称	加标量/(μ · mol L^{-1})	回收量/(μ · mol L^{-1})	相对标准偏差/%	回收率/%
功能饮料	0	2.60	2.9	—
	5	7.47	4.5	95.0
	10	12.52	2.6	96.9
钴胺素药片	0	8.63	2.1	—
	5	13.49	4.2	98.4
	10	18.62	7.2	99.9

本章还研究了 CBL 对 FCDs2 荧光淬灭的机理。图 4 - 17(a) 为 FCDs2 在 450 ~ 600 nm波长范围内的荧光激发光谱和它 CBL 的 UV - vis 吸收光谱,从图中可以看出两者具有较大范围的重叠。事实上,基于荧光光谱和吸收光的重叠所致的荧光淬灭通常被认为是通过内滤效应或荧光共振能量转移所引起的[25]。对于荧光共振能量转移过程,在激发态发生能量/电子转移会大大改变荧光团的寿命[57]。在此,FCDs2 显示出约 2.04 ns 的荧光寿命。添加 CBL 后,FCDs2 的荧光寿命为 2.07 ns,这种变化可以忽略不计[图 4 - 17(b)]。同时,加入 CBL 后,FCDs2 的荧光发射波长没有明显的蓝移或红移。因此,可以得出结论,CBL 对 FCDs2 的荧光淬灭机理是内滤效应。

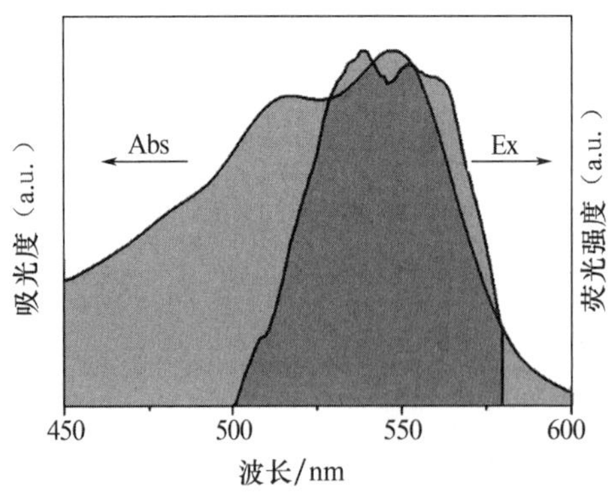

波长/nm

(a) CBL的UV-vis吸收光谱和FCDs2的荧光激发光谱

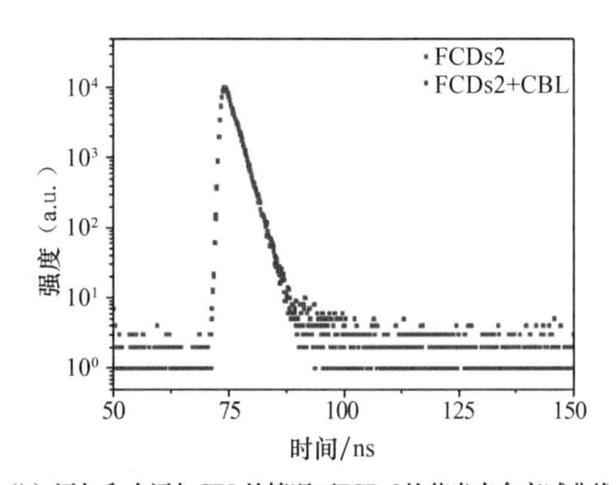

(b) 添加和未添加CBL的情况下FCDs2的荧光寿命衰减曲线

图 4 -17

4.4　小结

本章提出了一种基于氟掺杂的表面态控制法,在室温下制备了不同表面态 o - PD 衍生 CDs。室温制备方法克服了传统的高能耗制备方法的缺陷。研究发现,CDs 中氟掺杂含量的增加可以有效地诱导其荧光发射的红移。同时,制备的 CDs 还表现出了不同的固态和水性荧光特性,并证明了它们在 LFPs 显现和 CBL 传感中的实用性。氟掺杂诱导 CDs 产生的不同荧光行为不仅为碳纳米材料的表面态/结构调控提供了一个有效的方法,而且为室温下制备所需发射的 CDs 提供了一种简便的方案。

当然,本工作中对于 CDs 的荧光发射范围的调节还是有限的。因此,开发一种室温下对 CDs 荧光发射调节范围更大的方法将是未来研究的一个方向。

参考文献

[1] Xu X, Ray R, Gu Y, et al. Electrophoretic analysis and purification of fluorescent single - walled carbon nanotube fragments [J]. Journal of the American Chemical Society, 2004, 126 (40): 12736 - 12737.

[2] Sun Y - P, Zhou B, Lin Y, et al. Quantum - sized carbon dots for bright and colorful photoluminescence [J]. Journal of the American Chemical Society, 2006, 128: 7756 - 7757.

[3] Kou E, Yao Y, Yang X, et al. Regulation mechanisms of carbon dots in the development of lettuce and tomato [J]. ACS Sustainable Chemistry & Engineering, 2021, 9 (2): 944 - 953.

[4] Zhang Z, Zhang D, Shi C, et al. 3,4 - Hydroxypyridinone - modified carbon quantum dot as a highly sensitive and selective fluorescent probe for the rapid detection of uranyl ions [J]. Environmental Science: Nano, 2019, 6 (5): 1457 - 1465.

[5] Li T, Shi W, E S, et al. Green preparation of carbon dots with different surface states simultaneously at room temperature and their sensing applications [J]. Journal of Colloid and Interface Science, 2021, 591: 334 - 342.

[6] Wang X, Wang Y, Pan W, et al. Carbon - dot - based probe designed to detect intracellular pH in fungal cells for building its relationship with intracellular polysaccharide [J].

ACS Sustainable Chemistry & Engineering, 2021, 9 (10): 3718 – 3726.

[7] Gao X, Du C, Zhuang Z, et al. Carbon quantum dot – based nanoprobes for metal ion detection [J]. Journal of Materials Chemistry C, 2016, 4 (29): 6927 – 6945.

[8] Krysmann M J, Kelarakis A, Giannelis E P. Photoluminescent carbogenic nanoparticles directly derived from crude biomass [J]. Green Chemistry, 2012, 14 (11): 3141.

[9] Ding Z, Li F, Wen J, et al. Gram – scale synthesis of single – crystalline graphene quantum dots derived from lignin biomass [J]. Green Chemistry, 2018, 20 (6): 1383 – 1390.

[10] Liu J H, Li Y, He J H, et al. Polarity – sensitive polymer carbon dots prepared at room – temperature for monitoring the cell polarity dynamics during autophagy [J]. ACS Applied Materials & Interfaces, 2020, 12 (4): 4815 – 4820.

[11] Yan Y, Liu J H, Li R S, et al. Carbon dots synthesized at room temperature for detection of tetracycline hydrochloride [J]. Analytica Chimica Acta, 2019, 1063: 144 – 151.

[12] Lyu B, Li H – J, Xue F, et al. Facile, gram – scale and eco – friendly synthesis of multi – color graphene quantum dots by thermal – driven advanced oxidation process [J]. Chemical Engineering Journal, 2020, 388: 124285.

[13] Ding H, Yu S – B, Wei J – S, et al. Full – color light – emitting carbon dots with a surface – state – controlled luminescence mechanism [J]. ACS Nano, 2016, 10 (1): 484 – 491.

[14] Gao D, Zhang Y, Liu A, et al. Photoluminescence – tunable carbon dots from synergy effect of sulfur doping and water engineering [J]. Chemical Engineering Journal, 2020, 388: 124199.

[15] Dong Y, Pang H, Yang H B, et al. Carbon – based dots co – doped with nitrogen and sulfur for high quantum yield and excitation – independent emission [J]. Angewandte Chemie International Edition, 2013, 52 (30): 7800 – 7804.

[16] Wang M, Li M, Yu A, et al. Fluorescent nanomaterials for the development of latent fingerprints in forensic sciences [J]. Advanced Functional Materials, 2017, 27 (14): 1606243.

[17] Song W, Mao Z, Liu X, et al. Detection of protein deposition within latent fingerprints by surface – enhanced Raman spectroscopy imaging [J]. Nanoscale, 2012, 4 (7):

2333 - 2338.

[18] Day J S, Edwards H G M, Dobrowski S A, et al. The detection of drugs of abuse in fingerprints using Raman spectroscopy I: latent fingerprints [J]. Spectrochimica Acta Part A: Molecular and Biomolecular Spectroscopy, 2004, 60 (3): 563 - 568.

[19] Park J Y, Chung J W, Park S J, et al. Versatile fluorescent $CaGdAlO_4:Eu^{3+}$ red phosphor for latent fingerprints detection [J]. Journal of Alloys and Compounds, 2020, 824: 153994.

[20] Saif M, Alsayed N, Mbarek A, et al. Preparation and characterization of new photoluminescent nano - powder based on $Eu^{3+}:La_2Ti_2O_7$ and dispersed into silica matrix for latent fingerprint detection [J]. Journal of Molecular Structure, 2016, 1125: 763 - 771.

[21] Wang M, Li M, Yu A, et al. Rare earth fluorescent nanomaterials for enhanced development of latent fingerprints [J]. ACS Applied Materials & Interfaces, 2015, 7 (51): 28110 - 28115.

[22] Costa C V, Gama L I L M, Damasceno N O, et al. Bilayer systems based on conjugated polymers for fluorescence development of latent fingerprints on stainless steel [J]. Synthetic Metals, 2020, 262: 116347.

[23] Lai J A, Long Z, Qiu J, et al. Novel organic - inorganic hybrid powder $SrGa_{12}O_{19}:Mn^{2+}$ - ethyl cellulose for efficient latent fingerprint recognition via time - gated fluorescence [J]. RSC Advances, 2020, 10 (14): 8233 - 8243.

[24] Tang M, Zhu B, Qu Y, et al. Fluorescent silicon nanoparticles as dually emissive probes for copper(II) and for visualization of latent fingerprints [J]. Microchim Acta, 2019, 187 (1): 65.

[25] Du F, Cheng Z, Kremer M, et al. A label - free multifunctional nanosensor based on N - doped carbon nanodots for vitamin B_{12} and Co^{2+} detection, and bioimaging in living cells and zebrafish [J]. Journal of Materials Chemistry B, 2020, 8 (23): 5089 - 5095.

[26] Zhang L, Wang H, Hu Q, et al. Carbon quantum dots doped with phosphorus and nitrogen are a viable fluorescent nanoprobe for determination and cellular imaging of vitamin B_{12} and cobalt(II) [J]. Microchim Acta, 2019, 186 (8): 506.

[27] Jiang K, Sun S, Zhang L, et al. Bright - yellow - emissive N - doped carbon dots: Preparation, cellular imaging, and bifunctionalsensing, [J]. ACS Applied Materials &

Interfaces, 2015, 7: 23231 – 23238.

[28] Niu F, Ying Y – L, Hua X, et al. Electrochemically generated green – fluorescent N – doped carbon quantum dots for facile monitoring alkaline phosphatase activity based on the Fe^{3+} – mediating ON – OFF – ON – OFF fluorescence principle [J]. Carbon, 2018, 127: 340 – 348.

[29] Liu S G, Luo D, Li N, et al. Water – soluble nonconjugated polymer nanoparticles with strong fluorescence emission for selective and sensitive detection of nitro – explosive picric acid in aqueous medium [J]. ACS Applied Materials & Interfaces, 2016, 8 (33): 21700 – 21709.

[30] Zorn H, Li Q X. Trends in Food Enzymology [J]. Journal of Agricultural and Food Chemistry, 2017, 65 (1): 4 – 5.

[31] Jiang L, Ding H, Lu S, et al. Photoactivated fluorescence enhancement in F, N – doped carbon dots with piezochromic behavior [J]. Angewandte Chemie International Edition, 2020, 59: 9986 – 9991.

[32] Nie H, Li M, Li Q, et al. Carbon dots with continuously tunable full – color emission and their application in ratiometric pH sensing [J]. Chemistry of Materials, 2014, 26 (10): 3104 – 3112.

[33] Yarur F, Macairan J – R, Naccache R. Ratiometric detection of heavy metal ions using fluorescent carbon dots [J]. Environmental Science: Nano, 2019, 6 (4): 1121 – 1130.

[34] Yao K, Lv X, Zheng G, et al. Effects of carbon quantum dots on aquatic environments: Comparison of toxicity to organisms at different trophic levels [J]. Environmental Science & Technology, 2018, 52 (24): 14445 – 14451.

[35] Zuo G, Xie A, Li J, et al. Large emission red – shift of carbon dots by fluorine doping and their applications for red cell imaging and sensitive intracellular Ag^{+} detection [J]. The Journal of Physical Chemistry C, 2017, 121 (47): 26558 – 26565.

[36] Sun S, Zhang L, Jiang K, et al. Toward high – efficient red emissive carbon dots: Facile preparation, unique properties, and applications as multifunctional theranostic agents [J]. Chemistry of Materials, 2016, 28 (23): 8659 – 8668.

[37] Han L, Liu S G, Dong J X, et al. Facile synthesis of multicolor photoluminescent polymer carbon dots with surface – state energy gap – controlled emission [J]. Journal of Materials Chemistry C, 2017, 5 (41): 10785 – 10793.

[38] Dong Y, Li J, Zhang L. 3D hierarchical hollow microrod via in – situ growth 2D SnS nanoplates on MOF derived Co, N co – doped carbon rod for electrochemical sensing [J]. Sensors and Actuators B: Chemical, 2020, 303: 127208.

[39] Li Z, Wang D, Xu M, et al. Fluorine – containing graphene quantum dots with a high singlet oxygen generation applied for photodynamic therapy [J]. Journal of Materials Chemistry B, 2020, 8 (13): 2598 – 2606.

[40] Guo J, Ye S, Li H, et al. One – pot synthesized nitrogen – fluorine – codoped carbon quantum dots for ClO^- ions detection in water samples [J]. Dyes and Pigments, 2020, 175: 108178.

[41] Lu S, Sui L, Liu J, et al. Near – infrared photoluminescent polymer – carbon nanodots with two – photon fluorescence [J]. Advanced Materials, 2017, 29 (15): 1603443.

[42] Miao X, Yan X, Qu D, et al. Red emissive sulfur, nitrogen codoped carbon dots and their application in ion detection and theraonostics [J]. ACS Applied Materials & Interfaces, 2017, 9: 18549 – 18556.

[43] Song W, Duan W, Liu Y, et al. Ratiometric detection of intracellular lysine and pH with one – pot synthesized dual emissive carbon dots [J]. Analytical Chemistry 2017, 89 (24): 13626 – 13633.

[44] Kozák O, Sudolská M, Pramanik G, et al. Photoluminescent carbon nanostructures [J]. Chemistry of Materials, 2016, 28 (12): 4085 – 4128.

[45] Hu S, Trinchi A, Atkin P, et al. Tunable photoluminescence across the entire visible spectrum from carbon dots excited by white light [J]. Angewandte Chemie International Edition, 2015, 54 (10): 2970 – 2974.

[46] Zhan J, Geng B, Wu K, et al. A solvent – engineered molecule fusion strategy for rational synthesis of carbon quantum dots with multicolor bandgap fluorescence [J]. Carbon, 2018, 130: 153 – 163.

[47] Wang H – J, Hou W – Y, Yu T – T, et al. Facile microwave synthesis of carbon dots powder with enhanced solid – state fluorescence and its applications in rapid fingerprints detection and white – light – emitting diodes [J]. Dyes and Pigments, 2019, 170: 107623.

[48] Niu X, Song T, Xiong H. Large scale synthesis of red emissive carbon dots powder by solid state reaction for fingerprint identification [J]. Chinese Chemical Letters, 2021,

32 (6): 1953 – 1956.

[49] Wang J, Li Q, Zheng J, et al. N, B – codoping induces high – efficiency solid – state fluorescence and dual emission of yellow/orange carbon dots [J]. ACS Sustainable Chemistry & Engineering, 2021, 9 (5): 2224 – 2236.

[50] Xu J, Zhang B, Jia L, et al. Dual – mode, color – tunable, lanthanide – doped core – shell nanoarchitectures for anti – counterfeiting inks and latent fingerprint recognition [J]. ACS Applied Materials & Interfaces, 2019, 11 (38): 35294 – 35304.

[51] Gholami J, Manteghian M, Badiei A, et al. Label free detection of vitamin B_{12} based on fluorescence quenching of graphene oxide nanolayer [J]. Fullerenes, Nanotubes and Carbon Nanostructures, 2015, 23 (10): 878 – 884.

[52] Chen B B, Liu Z X, Deng W C, et al. A large – scale synthesis of photoluminescent carbon quantum dots: A self – exothermic reaction driving the formation of the nanocrystalline core at room temperature [J]. Green Chemistry, 2016, 18 (19): 5127 – 5132.

[53] Meng Y, Jiao Y, Zhang Y, et al. Facile synthesis of orange fluorescence multifunctional carbon dots for label – free detection of vitamin B_{12} and endogenous/exogenous peroxynitrite [J]. Journal of Hazardous Materials 2021, 408: 124422.

[54] Long Y, Zhang L, Yu Y, et al. Silicon nanoparticles synthesized using a microwave method and used as a label – free fluorescent probe for detection of VB_{12} [J]. Luminescence, 2019, 34 (6): 544 – 552.

[55] Wongyai S. Determination of vitamin B in multivitamin tablets by multimode 12 high – performance liquid chromatography [J]. Journal of Chromatography A, 2000, 870: 217 – 220.

[56] Tomčik P, Banks C E, Davies T J, et al. A self – catalytic carbon paste electrode for the detection of vitamin B_{12} [J]. Analytical Chemistry 2004, 76: 161 – 165.

[57] Lu S, Wang S, Zhao J, et al. Fluorescence light – up biosensor for microRNA based on the distance – dependent photoinduced electron transfer [J]. Analytical Chemistry, 2017, 89 (16): 8429 – 8436.

第5章

溶剂热法制备黄光氮掺杂碳点及其应用

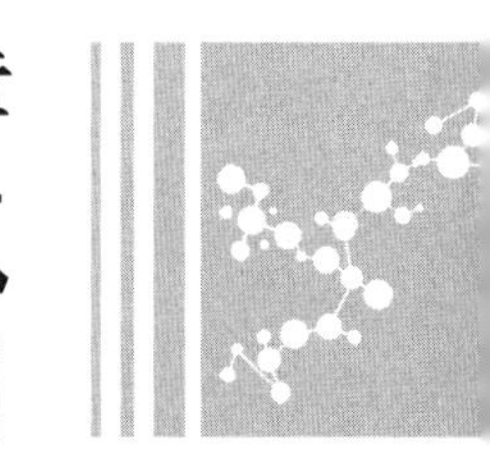

5.1 引言

近年来,CDs作为荧光纳米材料家族的新成员,因其杰出的光学性能、优异的水分散性、良好的生物相容性等诸多优点而备受关注[1-9]。基于上述特点,CDs已被广泛应用于传感、生物成像、光催化等领域[10-16]。例如,Xiong等人基于水热法和色谱分离技术制备了长波长发射的CDs,他们发现CDs的荧光红移取决于表面基团,简单来说CDs的荧光红移是带隙随着表面氧结构的增加而减小所引起的[17]。此外,由于CDs具有突出的荧光发射特性和较低的细胞毒性,还可以用于细胞成像。Wang等报道了利用间氨基苯酚作为碳前驱体,通过溶剂热法获得黄绿色发射CDs(λ_{em} = 520 nm)[18]。虽然这些方法可以制备不同性质和用途的CDs,但它们都有一个共同的缺点,即纯化过程烦琐。此外,为了改善和调整CDs的性质,杂原子掺杂作为一种简单的方法已被广泛使用[19-24]。通过杂原子掺杂,CDs可以被赋予一种不同的表面态和独特的特性。例如Sarkar等人通过比较未掺杂和掺杂磷的CDs,证实了磷掺杂对CDs的荧光发射有一定程度影响。他们发现磷掺杂可以引起最大50 nm的红移,并提高量子产率[25]。Prasad等人利用N,N-二甲基甲酰胺和磷酸制备了磷、氮共掺杂CDs[26]。由于杂原子功能化,CDs在碱性介质的氧化还原反应中表现出优异的荧光特性、高的电催化活性。由此可见,杂原子掺杂对CDs的荧光性质有显著影响,是实现CDs功能化的有效方法。

姜黄素的历史悠久,早在千年前就已经以药物身份出现在中医复方中药和古印度的阿育吠陀疗法中[27]。姜黄素提取自中药姜黄根茎,分子式为$C_{21}H_{20}O_6$,是一种亲脂性生物活性多酚。中药姜黄根茎可提取1%~6%的姜黄素类化合物,其中姜黄素占姜黄素类似物的60%~70%,是最具生物活性的组分[27]。在不同酸碱度的化学环境下,姜黄素含有7个碳原子的碳链上具有的2个酮基可以进行双酮-烯醇互变,因此姜黄素在人体生理状态pH值下的化学结构并不稳定。姜黄素的主要功能体现在它的官能团:酚基和二酮结构。这两种具有活性的官能团介导姜黄素的供氢反应、迈克尔加成反应以及一系列水解和酶促反应。姜黄素结构中的酚羟基最易发生递氢反应,继而代谢成为苯氧基,发挥抗氧化的特性,清除由分子氧化剂和自由基氧化剂组成的活性氧。其次,姜黄素可作为亲核试剂与强亲电子试剂发生迈克尔加成反应,发挥抗癌细胞的细胞毒作用,即抗肿瘤活性。此外,姜黄素的毒副作用小并且具有广泛的药理活性,可以通过多种作用机理达到抗肿瘤、抗炎症、抗氧化应激、抗感染和调节脂质代谢等功能。姜黄素可以通过调节肿瘤细胞的分裂周期以抑制肿瘤细胞的增殖,还可通过抑制肿瘤细胞的迁移和浸润以达到抗肿瘤的作用。姜黄素不仅具有广谱抗肿瘤作用,还具有针对多系统中的以炎症、氧化应激等病因诱导的疾病的治疗效果。姜黄素凭借其极低的毒副作用和广泛的生物活性,为临床疾病的治疗提供了新的思路。姜黄素多年来一直用于中药和膳食补充剂中[28-30]。医学研究表明,姜黄素对人体健康有益,因此姜黄素在临床上被广泛用于治疗糖尿病、过敏、抑郁症、关节炎和阿尔茨海默病[31,32]。然而,随着对姜黄素研究的不断增加,科研人员发现高剂量姜黄素很可能对 DNA 产生促氧化活性,导致细胞间 ATP 水平下降和细胞坏死[33]。因此,姜黄素的定量检测对食品安全和人体健康尤为重要。到目前为止,姜黄素的分析方法主要有高效液相色谱法、紫外可见光谱法、薄层色谱法和荧光光谱法[34]。相比之下,荧光光谱法因其操作简便、消耗低、选择性高、响应快等优点而备受关注。因此,开发一种高效、简便的姜黄素荧光检测方法是十分必要和重要的。

近年来,信息安全与防伪受到了人们的广泛关注。荧光油墨的开发则是信息安全和防伪的基础[35]。荧光油墨是利用荧光颜料制成的具有荧光特性的油墨,荧光油墨在特定的激发光照射下会反馈出特殊信息[36,37]。在防伪领域,因其隐蔽性好、成本低廉、识别仪器普及、色彩鲜艳等特点而被广泛使用[38,39]。此外,在证券防伪、医药防伪、香烟防伪等包装上都能看到荧光油墨的应用。但随着科学技术的不断发展,研究人员发现目前的商用荧光油墨普遍存在荧光粉颗粒形状不规整、粒径分布范围大、在可见光条件下不透明等问题。因此,研发新型防伪材料和高效荧光油墨,具有重要的现实意义。CDs 由于具有成本低、光学性能优异等特点,在信息加密防伪方面有着广泛的应用[35,39-41]。

溶剂热法是在水热法的基础上发展起来的,指密闭体系如高压釜内,以有机物或非水溶媒为溶剂,在一定的温度和溶液的自生压力下,原始混合物进行反应的一种制备方法。该方法所用碳源广泛,从糖类、有机酸到生活中的果汁都可以作为碳源。因此,溶剂热法是目前制备 CDs 最常用的方法之一。本研究中使用简单、低成本的溶剂热策略,以 o-PD 和乙醇为原料,无须进一步修饰,制备黄色发射氮掺杂 CDs(YNCDs)。姜黄素能有效地淬灭 YNCDs 的荧光发射,且具有良好的选择性。在此基础上,研制了一种基于纳米探针的姜黄素检测传感器。结果表明,该方法具有较高的选择性和灵敏度,可用于实际样品中姜黄素的检测。此外,由于具有杰出的荧光特性、出色的稳定性和日光下的透明度,YNCDs 还可以作为隐形安全油墨用于防伪和信息加密。图 5-1 中介绍了 YNCDs 的制备和传感应用。

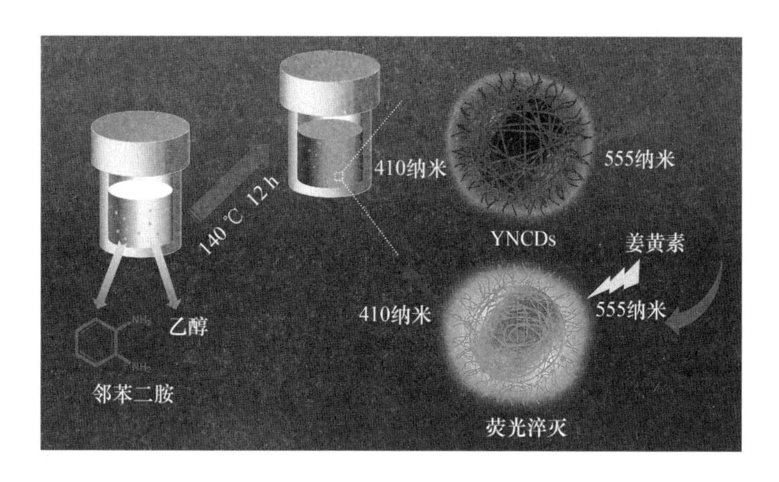

图 5－1　YNCDs 的制备及其传感应用的示意图

5.2　实验部分

5.2.1　实验仪器

KQ－100B/800KDE 型超声波清洗器(中国昆山市超声仪器有限公司);

BSA22AS 单盘型分析电子天平(中国北京赛多利斯仪器有限公司);

TG16－WS 台式高速离心机(中国湘仪实验室仪器开发有限公司);

2XZ－2 型真空泵(中国临海市谭式真空设备有限公司);

PB－10 标准型 pH 计(中国北京赛多利斯仪器有限公司);

Lambda Bio20 紫外可见光分光光度计(美国珀金埃尔默仪器有限公司);

F－7000 荧光分光光度计(日本日立公司);

JEM－2100 透射电子显微镜(日本电子株式会社);

Bruker Dimension icon 原子力显微镜(德国布鲁克公司);

DZF－6020 型真空干燥箱(中国上海精宏实验设备有限公司);

One Spectra 红外光谱仪(美国珀金埃尔默仪器有限公司);

DHG－9037A 电热恒温干燥箱(中国上海精宏实验设备有限公司);

YE5A44 型手动可调式移液器(中国上海大龙医疗设备有限公司);

CHI660E 电化学工作站(中国上海晨华仪器有限公司);

EscaLab 250Xi X 射线光电子能谱分析仪(美国赛默飞世尔公司);

Malvern Nano－ZS 粒度仪(英国马尔文公司);

5.2.2 实验试剂

本章所使用化学试剂和品牌如下:

O－PD、Cys、葡萄糖、抗坏血酸(AA)、谷胱甘肽(GSH)以及本章使用的其他氨基酸购自阿拉丁化学有限公司(中国上海)。乙醇、硝酸银($AgNO_3$)以及本章使用的其他无机盐购自国药集团化学试剂有限公司(中国上海)。

除特别声明外,所有试剂皆为分析纯且未经任何前处理。实验用水为二次去离子水(18 MΩ cm)。

5.2.3 YNCDs 的制备方法

将 0.2 g o－PD 加入到 20 mL 乙醇中,超声 10 分钟后,将溶液转移到 50 mL 的聚四氟乙烯内胆高压釜中,在 140 ℃加热 12 小时。冷却至室温后,将所得溶液通过 0.22 μm 过滤膜去除大颗粒。之后,在 30 ℃旋转蒸发得到粗产物。然后,将产物溶于去离子水中,通过纤维素酯膜(MWCO:500～1000 Da)透析 48 小时除去未反应的物质。然后收集溶液并冷冻干燥获得 CDs,将其命名为 YNCDs。

5.2.4 YNCDs 的表征方法

通过循环伏安法测试了 CDs 的电化学性质。其中以玻璃碳电极作为工作电极,铂丝作为辅助电极,Ag/AgCl 作为参比电极。在 0.1 $mol \cdot L^{-1}$ 四正丁基六氟磷酸铵的乙腈溶液中测试了 CDs 样品的循环伏安曲线。CDs 的 HOMO 和 LUMO 能级根据以下公式计算得出:

$$E_{LUMO} = -e(E_{red} + 4.4) \tag{5-1}$$

$$E_{HOMO} = E_{LUMO} - E_g \tag{5-2}$$

E_{LUMO}代表最低未占据分子轨道的能级,E_{HOMO}代表最高占据分子轨道的能级,E_{red}代表还原电位的起始值,E_g 代表带隙。

此外,本章通过 U－3900 紫外可见分光光度计使用 1 cm 光程的比色皿,在扫描间隔为 1 nm 条件下测得紫外－可见(UV－vis)吸收光谱。荧光光谱是通过 F－7000 荧光光谱仪使用 1 cm 光程的比色皿测得,激发和发射狭缝均设置为 10 nm。傅里叶红外(FT－IR)光谱使用 Nicolet－6700 红外光谱仪采用溴化钾压片法测定并记录 1 000～4 000 cm^{-1}的数据。X 射线光电子能谱(XPS)在配备 Al Kα 280.00 eV 激发光源的 ES-

CALAB 250 表面分析系统上进行测得。透射电子显微镜(TEM)图像通过 JEM-2100 高分辨透射电子显微镜测得,加速电压为 200 kV。

5.2.5 YNCDs 的稳定性测试

稳定性测试:取一定量的 YNCDs 溶液,分别储存 0~7 天。在 YNCDs 最佳激发波长下,测试上述溶液荧光发射光谱。每组样品测试 3 次。

离子强度稳定性测试:取 6 组 YNCDs 溶液,分别加入不同浓度的 NaCl,使 YNCDs 溶液中 Na^+ 浓度分别为:0、0.2、0.4、0.6、0.8、1.0 mol · L^{-1}。在 YNCDs 最佳激发波长下,测试上述溶液荧光发射光谱。每组样品测试 3 次。

5.2.6 YNCDs 用于姜黄素定量检测

通过以下步骤用 YNCDs 作为荧光探针进行姜黄素测定。首先制备姜黄素的乙醇原液,然后用磷酸缓冲溶液(pH=7.0)稀释 10 倍。将 1.0 mL 不同浓度的姜黄素溶液与 1.0 mL YNCDs 溶液(20.0 μg/mL,在磷酸缓冲溶液中制备)混合。之后,在 410 nm 的激发下测试荧光发射光谱。以姜黄素浓度为横坐标,荧光淬灭率$(F_0-F)/F_0$为纵坐标绘图,其中 F_0 和 F 分别代表添加姜黄素前后 YNCDs 的荧光强度。此外,在相同条件下将姜黄素替换为干扰物研究了 YNCDs 的选择性。所有测试均重复 3 次。

本研究以 YNCDs 作为探针通过加标回收法对咖喱粉、人体血清、尿液中的姜黄素进行了定量分析。咖喱粉可以在当地超市买到。先将 50.0 mg 咖喱粉样品溶于 10.0 mL乙醇溶液中,超声处理 10 分钟,然后将上述溶液离心 10 分钟,提取上清液过滤得到棕黄色姜黄素样品溶液。然后将姜黄素样品溶液用磷酸缓冲溶液稀释 10 倍,4 ℃保存,用于后续的真实样品测试。

所有的人体血清和尿液样本都来自健康志愿者。在血清样品中加入甲醇沉淀蛋白质,样品以 4 000 r · min^{-1}离心 10 分钟,通过 0.45 μm 膜过滤。然后,每个滤液用磷酸缓冲溶液稀释 10 倍,供进一步使用。

5.3 结果与讨论

5.3.1 YNCDs 的制备与表征

本研究以 o-PD 为碳源,乙醇为溶剂,无须特殊设备和氧化剂,在高温下制备 CDs。

研究发现,o－PD 上的 $-NH_2$ 基团具有较高的活性,使 o－PD 在高温高压下容易发生氧化或聚合反应,生成 o－PD 二聚体、o－PD 三聚体和2,3－二硝基苯胺等[12]。随着反应的持续进行,o－PD 的聚合反应不断形成长链聚 o－PD,这些聚合物可以纠缠在一起,最终形成 CDs。

为了研究制备 CDs 的结构和成分对其表面态的影响,研究中采用透射电镜(TEM)、傅里叶变换红外(FT－IR)和 X 射线光电子能谱(XPS)对制备的 CDs 进行了表征。TEM 是把经加速和聚集的电子束投射到非常薄的样品上,电子与样品中的原子碰撞而改变方向,从而产生立体角散射。散射角的大小与样品的密度、厚度相关,因此可以形成明暗不同的影像,影像将在放大、聚焦后在成像器件(如荧光屏、胶片、以及感光耦合组件)上显示出来。TEM 可以看到在光学显微镜下无法看清的小于0.2 μm 的细微结构,这些结构称为亚显微结构或超微结构。因此,TEM 被公认为是确定 CDs 形状、形态和大小的有效手段。图5－2(a)为 CDs 的 TEM 图,图中显示了它们由近似球形和分散良好的纳米颗粒组成。图5－2(a)中的插图显示,YNCDs 具有相对狭窄的尺寸分布,尺寸在3.0～6.5 nm 范围内,平均直径为5.3 nm,高分辨率 TEM(HRTEM)图像显示,YNCDs 具有分辨率较高的晶格条纹,间距为0.21 nm。

分子中组成化学键或官能团的原子处于不断振动的状态,其振动频率与红外光的振动频率相当。当用红外光照射分子时,分子中的化学键或官能团可发生振动吸收,不同的化学键或官能团吸收频率不同,在 FT－IR 上将处于不同位置,从而可获得分子中含有何种化学键或官能团的信息。为了研究 YNCDs 的表面态和组成,本研究对 YNCDs 进行了 FT－IR 表征,从图5－1(b)可以证实 O－H 在3 373 cm^{-1}处的伸缩振动、N－H 在3 172 cm^{-1}处的伸缩振动、C＝N 在1 635 cm^{-1}处的伸缩振动、C＝C 在1 506 cm^{-1}处的伸缩振动、C－N 在1 357 cm^{-1}处的伸缩振动、C－O 在1 206 和1 087 cm^{-1}处的伸缩振动的特征吸收带,上述结果表明 YNCDs 结构中聚芳族结构的形成和含氮基团的存在[12,42－44]。

XPS 的原理是用 X 射线去辐射样品,使原子或分子的芯电子或价电子受激发射出来。被光子激发出来的电子称为光电子。可以测量光电子的能量,以光电子的动能/束缚能为横坐标,相对强度为纵坐标可做出光电子能谱图,从而获得试样有关信息。XPS 光谱进一步分析了 YNCDs 的元素组成,发现 YNCDs 的 C 1s、N 1s 和 O 1s 峰分别位于284 eV、399 eV 和 532 eV 处,相对元素含量分别为73.0%、13.2%和13.8%图5－2(c)[12]。在高分辨率 XPS 光谱中,C 1s 图5－2(d)可以被分成为5个结合能峰,包括:284.1 eV 时的 C＝C/C－C、285.2 eV 时的 C－N、286.0 eV 时的 C－O、287.5 eV 时的 C＝N 和289.1 eV 时的 CO－NH[45]。N 1s 谱图图5－2(e)显示在398.5 eV、

399.2 eV、400.1 eV 和 401.1 eV 处有 4 个峰,分别属于吡啶氮,氨基氮,吡咯氮和石墨氮[17]。O 1s 谱图图 5－2(f)分别在 531.9 eV 和 532.8 eV 处包含两个峰对应 C＝O 和 C－O[17]。上述 XPS 结果与 FT－IR 光谱结果一致,进一步证明了 YNCDs 中这些官能团的存在,表明 YNCDs 成功制备。

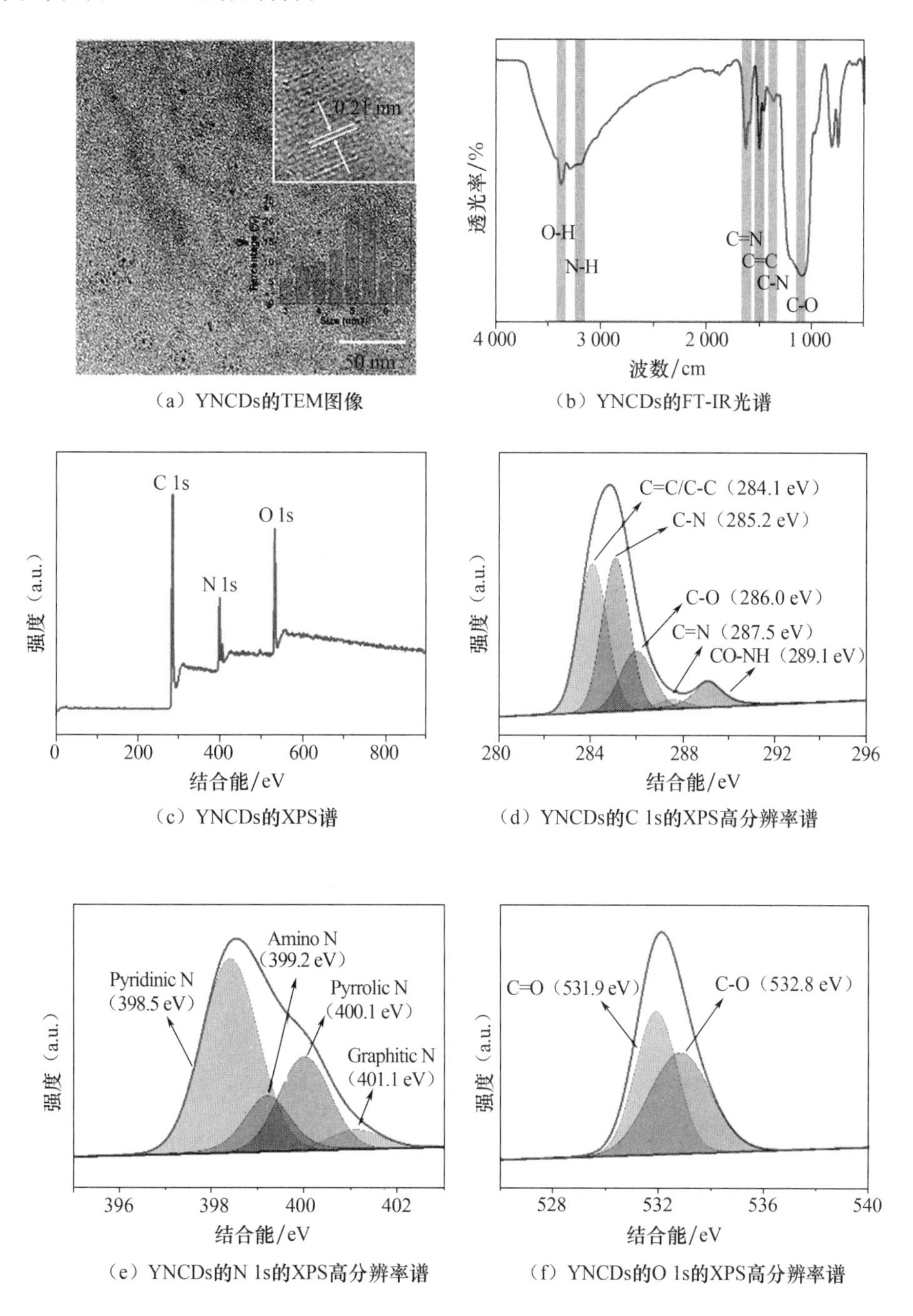

（a）YNCDs的TEM图像

（b）YNCDs的FT-IR光谱

（c）YNCDs的XPS谱

（d）YNCDs的C 1s的XPS高分辨率谱

（e）YNCDs的N 1s的XPS高分辨率谱

（f）YNCDs的O 1s的XPS高分辨率谱

图 5－2

5.3.2 YNCDs 的光学性质

在本章的研究中,利用荧光光谱和 UV - Vis 技术对制备的 YNCDs 进行了表征,并计算了 CIE 颜色坐标。在有机化合物分子中有形成单键的 σ 电子、有形成双键的 π 电子、有未成键的孤对 n 电子。当分子吸收一定能量的辐射能时,这些电子就会跃迁到较高的能级,此时电子所占的轨道称为反键轨道,而这种电子跃迁同内部的结构有密切的关系,这就产生了 UV - Vis 光谱。在图 5 - 3(a)中,YNCDs 在 200 ~ 360 nm 高能区中均表现出了 C = C 引起的 $\pi-\pi^*$ 跃迁和 C = O/C = N 引起的 $n-\pi^*$ 跃迁的吸收带。除了这些吸收带,在 360 ~ 700 nm 的低能区也观察到了吸收带,这归因于共轭 C = N 的 $n-\pi^*$ 跃迁[2,46]。同时,当 YNCDs 被 410 nm 的激发光照射时,在 555 nm 处检测到最强的荧光发射峰。进一步地,YNCDs 的荧光激发光谱与吸收光谱重叠,说明氮氧相关基团的结构促成了 YNCD 的黄色荧光发射。在 3D 荧光图图 5 - 3(b)中,显示了 YNCDs 存在单个发射中心,发射最大值在 555 nm 处。图 5 - 3(c)随着发射波长从 380 到 460 nm 的变化,发光强度先增大后减小,在发射波长 410 nm 时达到最大值。但与传统 CDs 相比,YNCDs 展现了不同的发射特性,即发射位置不随激发波长的变化而发生蓝/红位移[47-49]。产生这个结果通常归因于表面态发射[45,50]。用 CIE 颜色坐标进一步解释了 YNCDs 的荧光发射行为,如图 5 - 3(d)所示,YNCDs 的坐标在黄色区域(0. 3829, 0. 5697)。

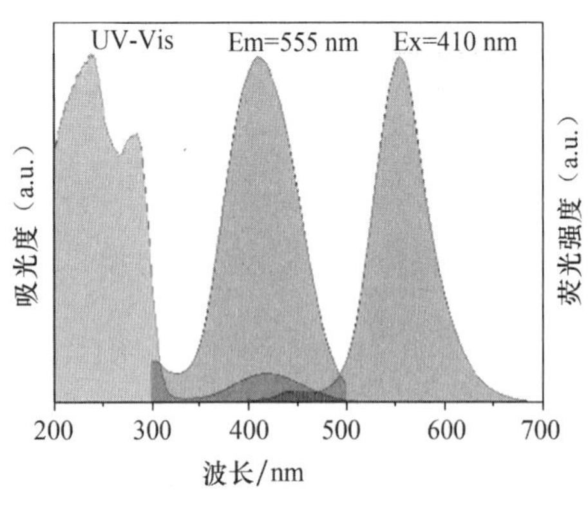

(a) YNCDs的UV-Vis、荧光激发和发射光谱

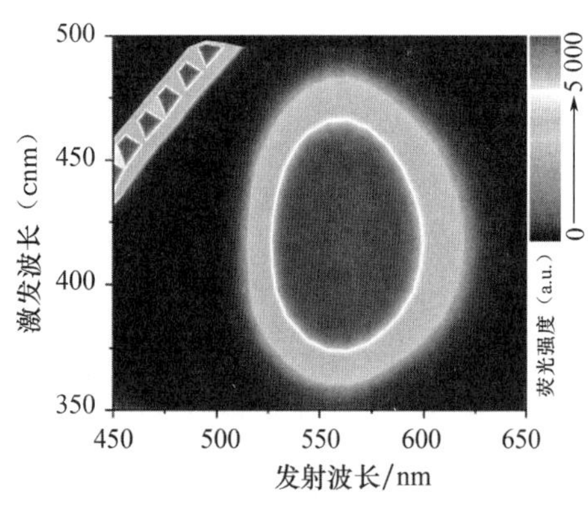

(b) YNCDs三维荧光图

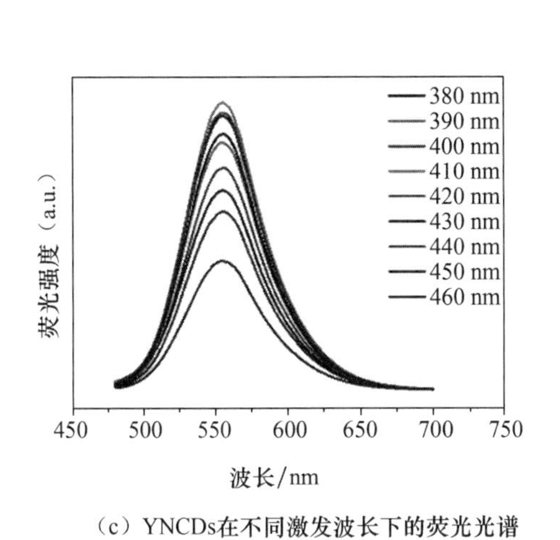

（c）YNCDs在不同激发波长下的荧光光谱

（d）YNCDs的CIE颜色坐标

图 5－3

通常来说，CDs 的荧光特性在极端条件下容易发生变化，因此其稳定性在实际应用中起着关键作用。如图 5－4 和图 5－5 所示，经过 7 天和不同盐溶液浓度的保存，YNCDs 的归一化荧光强度保持不变，表明具有良好的光稳定性和结构稳定性。

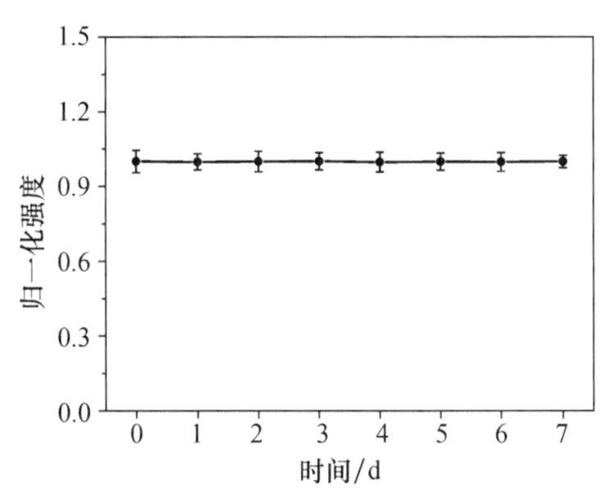

图 5－4　新鲜制备和储存(7 d)的 YNCDs 的荧光强度比较

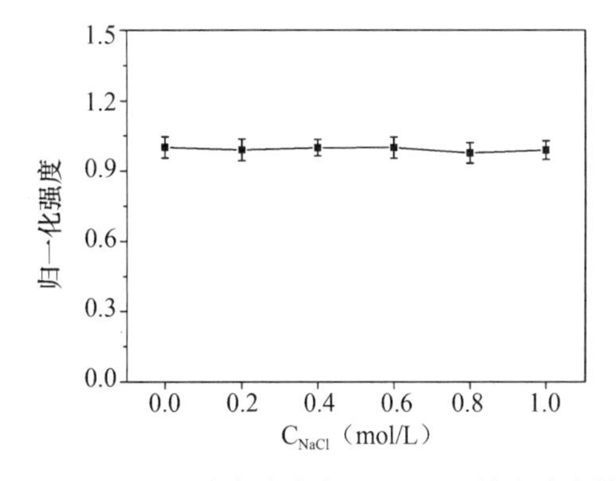

图 5－5　不同盐溶液浓度下 YNCDs 的光稳定性

5.3.3 基于 YNCDs 的姜黄素荧光检测

随着对姜黄素研究的深入,人们发现高剂量姜黄素很可能对 DNA 产生促氧化活性,导致细胞间 ATP 水平下降和细胞坏死[33]。因此,姜黄素的定量检测对食品安全和人体健康尤为重要。姜黄素可以有效地淬灭 YNCDs 的荧光发射,这为我们建立姜黄素的荧光纳米探针提供了可能。进一步地,本章通过优化条件探讨了开发基于 YNCDs 检测姜黄素的可行性。图 5-6(a)显示了 YNCDs 对姜黄素在 0~90.0 μmol · L^{-1}范围内的荧光响应性。随着姜黄素的加入,在 555 nm 处 YNCDs 的荧光强度逐渐降低,说明纳米探针对姜黄素浓度非常敏感[图 5-6(a)]。淬灭率($(F_0-F)/F_0$)与姜黄素浓度的关系如图 5-6(b)所示,当姜黄素浓度在 0.04~30.0 μmol · L^{-1},淬灭率($(F_0-F)/F_0$)与姜黄素浓度之间存在良好的线性关系,线性回归方程为$(F_0-F)/F_0 = 0.132C + 0.002\ 55$,$R^2$ 为 0.997 见图 5-6(b),计算出的检出限(3S/N)为 0.01 μmol · L^{-1},与表 5-1 所示的已报道的姜黄素检测文献相比,本章的结果令人满意。

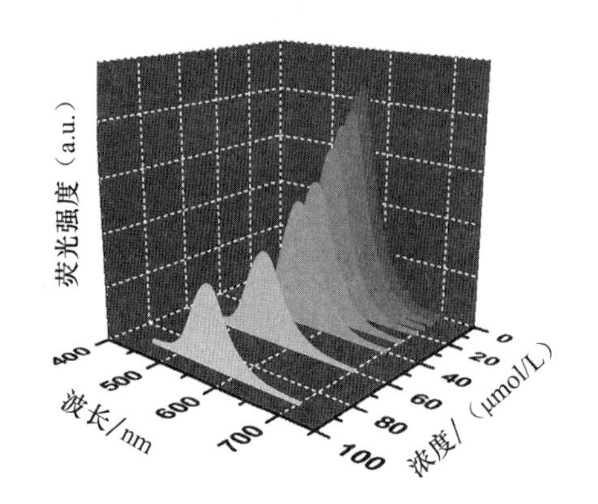

(a)不同浓度(0~90.0 μmol · L^{-1})姜黄素下YNCDs的荧光光谱

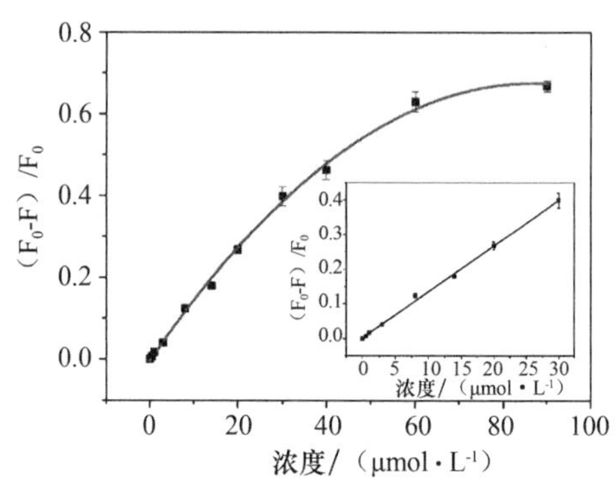

(b)YNCDs的(F_0-F)/F_0与姜黄素浓度的关系(插图姜黄素浓度在0.04~30.0 μmol · L^{-1}范围内的线性相关图)

图 5-6

表 5-1 检测姜黄素分析方法的比较

探针	检出限/(μmol · L^{-1})	线性范围/(μmol · L^{-1})	参考文献
N-CDs	0.054	0.54~22.0	[28]
BNCDs	0.065	0.2~12.5	[29]
p-CDs	0.133	0.4~45.0	[33]

续表

探针	检出限/(μmol·L^{-1})	线性范围/(μmol·L^{-1})	参考文献
Urazole-AuNCs	0.072	1.0~60.0	[30]
Ag-g-C_3N_4	0.038	0.01~2.0	[31]
YNCDs	0.01	0.04~30.0	本工作

选择性是评价探针特异性的关键参数。本章研究了几种常见金属离子、葡萄糖和氨基酸等干扰物质对姜黄素测定的影响。如图 5-7 所示,姜黄素(30 μmol·L^{-1})对 YNCDs 的荧光淬灭效果比其他物质(300 μmol·L^{-1})明显。简单来说,各种干扰物质的共存对检测系统的影响可以忽略不计,该实验证实了 YNCDs 探针对姜黄素具有优异的选择性。

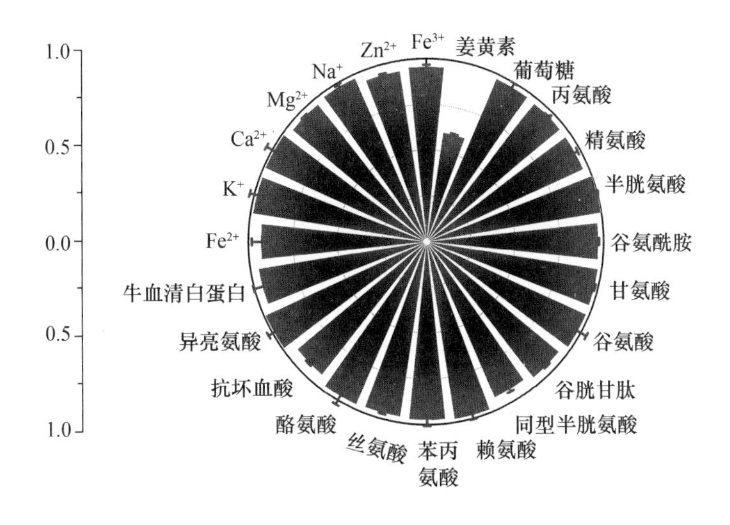

图 5-7　利用 YNCDs 对姜黄素进行选择性检测

为了验证 YNCDs 荧光探针的实用性和可行性,实验中选取咖喱粉、人血清、人尿液和饮用水作为姜黄素定量分析的实际样品。本研究采用加标回收法,在处理后的样品中加入已知量的姜黄素标准溶液(1.0 和 10.0 μmol·L^{-1}),结果如表 5-2 所示。结果表明,回收率在 96.5%~108.0%之间,相对标准偏差(RSD)<3.8%(n=3),证明该方法在实际应用中具有较高的准确性。

表 5－2　实际样品中姜黄素的测定结果

样品名称	加标量/(μmol·L^{-1})	回收量/(μmol·L^{-1})	回收率/%	相对标准偏差/%
咖喱粉	0.0	0.84	—	2.5
	1.0	1.93	108.0	3.2
	10.0	10.75	99.1	2.9
人尿液	1.0	0.98	98.0	3.2
	10.0	10.34	103.4	2.8
人血清	1.0	1.04	104.0	3.8
	10.0	9.67	96.7	3.1
饮用水	1.0	0.97	97.0	3.4
	10.0	9.65	96.5	3.2

5.3.4　姜黄素检测的淬灭机理

通常来说，姜黄素对 CDs 的荧光淬灭过程与内滤效应(IFE)有关[51]。本章中为了证实姜黄素对 YNCDs 的淬灭机理是 IFE，对 YNCDs 和姜黄素进行了光学性能测试。如图 5－8(a)所示，YNCDs 的荧光激发光谱与姜黄素的吸收光谱几乎完全重叠，荧光共振能量转移(FRET)、光致电子转移(PET)和 IFE3 种淬灭机理均依赖于荧光团的激发或发射区域与淬灭剂的吸收区域一定程度的重叠。事实上，在 FRET 过程中激发态能量/电子转移的发生会显著改变荧光团的寿命[13]。在图 5－8(b)中，YNCDs 的荧光寿命约为 2.38 ns，姜黄素的加入几乎不会对 YNCDs 的荧光寿命(2.37 ns)造成任何改变。这个结果说明 FRET 并不是姜黄素对 YNCDs 的荧光淬灭的机理。

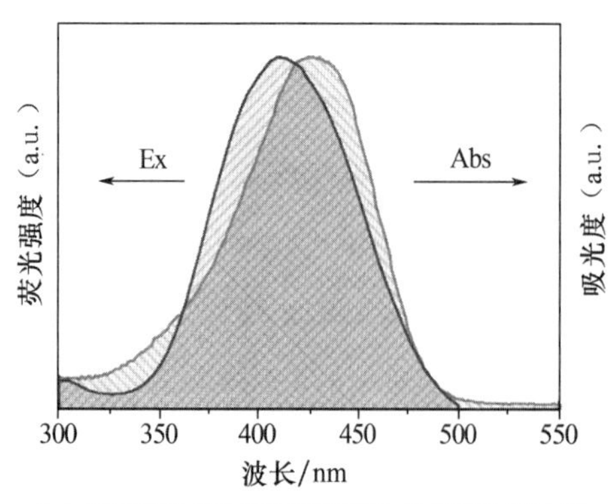

(a) 姜黄素的吸收光谱和YNCDs的荧光发射光谱

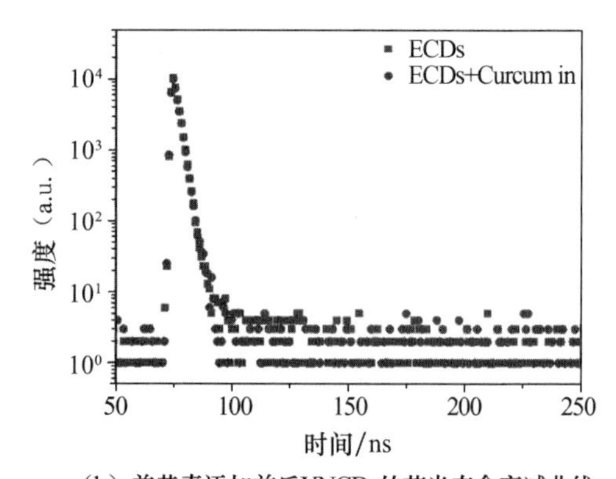

(b) 姜黄素添加前后YNCDs的荧光寿命衰减曲线

图 5－8

为了排除 PET 引起 YNCDs 的荧光猝灭可能性，本章采用循环伏安法（CV）研究了 YNCDs 的最高占据轨道（HOMO）和最低未占据轨道（LUMO）能级。电极的制备和测试过程如下，CV 使用标准的三电极体系进行测试，该体系含有饱和 CDs 和 0.1 mol · L^{-1} 的 $TBAPF_6$ 乙腈溶液作为支撑电解质。根据以下公式计算 CDs 的 eV 中的 HOMO 和 LUMO 能级：

$$E_{LUMO} = -e(E_{red} + 4.4) \tag{5-1}$$

$$E_{HOMO} = E_{LUMO} - E_g \tag{5-2}$$

其中，E_{LUMO}是 LUMO 的能级，E_{HOMO}是 HOMO 的能级，E_{red}是还原电位

得到的 YNCDs 的 E_{red} 为 -0.51 V［图 5-9(a)］。根据式(5-1)估计 YNCDs 的 E_{LUMO}为 -3.89 eV。根据 YNCDs 的紫外吸收光谱［图 5-9(b)］，计算出 E_g 为 2.56 eV。根据公式(5-2)估计 YNCDs 的 E_{HOMO}为 -6.45 eV。根据文献报道姜黄素的 E_{HOMO}和 E_{LUMO}分别为 -5.06 eV 和 -3.16 eV[28]。如图 5-9(c)所示，YNCDs 的激发电子并没有从其LUMO转移到姜黄素的 LUMO。也就是说，PET 不能引起 YNCDs 的荧光猝灭。综上所述，YNCDs 的荧光激发光谱与姜黄素的吸收光谱的大面积重叠，以及在有/无姜黄素情况下 YNCDs 的荧光寿命未发生变化，很好地说明了 YNCDs 的荧光猝灭应归因于 IFE。

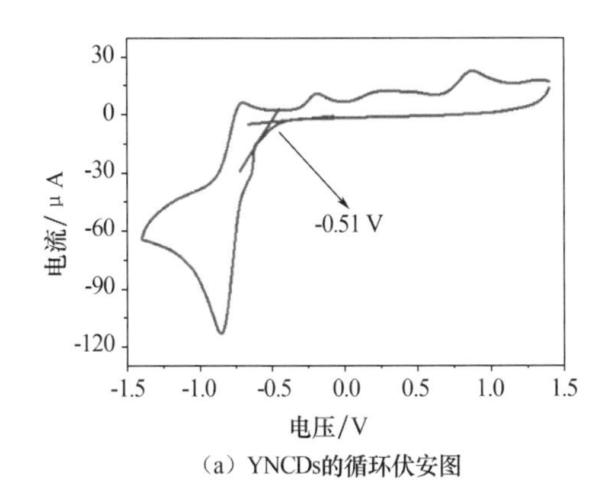

(a) YNCDs的循环伏安图

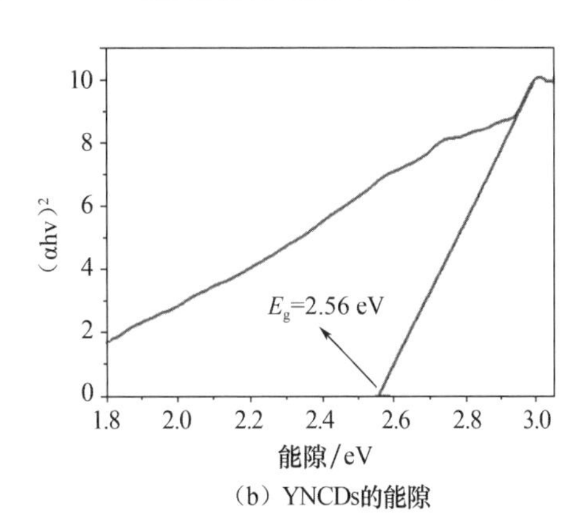

(b) YNCDs的能隙

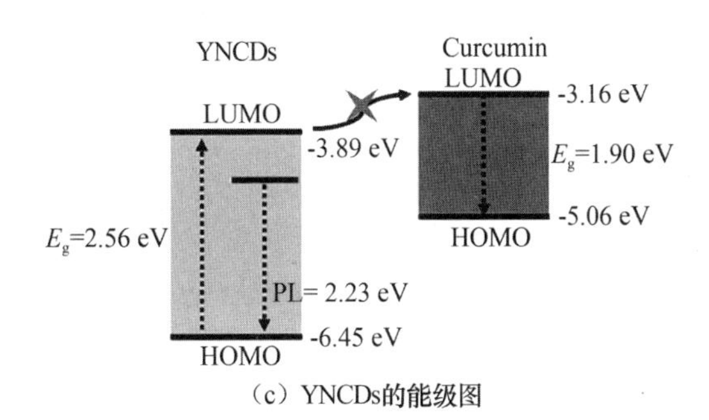

(c) YNCDs的能级图

图 5-9

5.3.5 荧光防伪墨水

CDs 不但具有荧光性能，还兼具亲水/油性的特点，同时其对书写材质的包容性强，因此 CDs 可作为荧光墨水用于书写[39,52,53]。2018 年，Bandi 等人利用富含纤维素和芳香族官能团的废弃烟头作为原料，制备的 CDs 量子产率高达 26%，研究发现该 CDs 具有高稳定发光性能并在室温下保存 30 天仍可书写，这使该 CDs 在荧光墨水领域具有良好的应用前景[54]。2018 年，Paul 等人利用氨基酸修饰 CDs 制备出了两亲性荧光凝胶，制备的凝胶即使在干燥固体状态下也能显示出显著的绿色荧光，因此这种水凝胶可以用作荧光墨水[55]。在本章中，由于 YNCDs 独特的光学特性，它展示了多种应用的可能。除了检测姜黄素外，具有良好水溶性和优异光学性能的 YNCDs 还可以作为荧光防伪油墨用于信息存储和高级防伪。如图 5－10 和图 5－11 所示，在日光下图案是不可见的（图 5－10），但在紫外线下图案会出现（图 5－10 和图 5－11），这保证了负载信息的安全。同时，在室温下保存数月后，标记图案仍保持良好的稳定性。与传统油墨相比，基于 YNCDs 的油墨具有书写清晰、永久保存、不污染环境等优点。

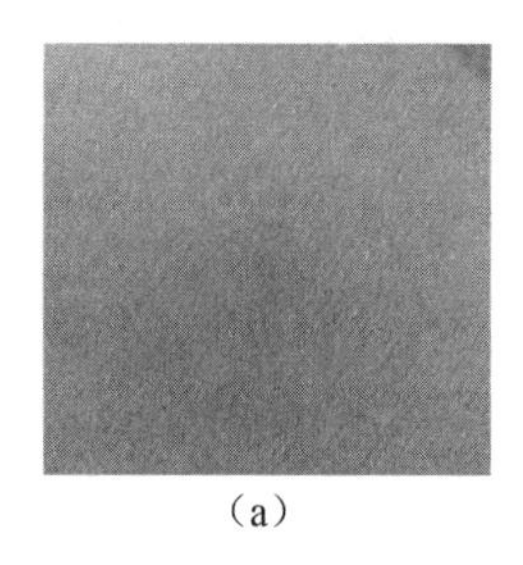
（a）

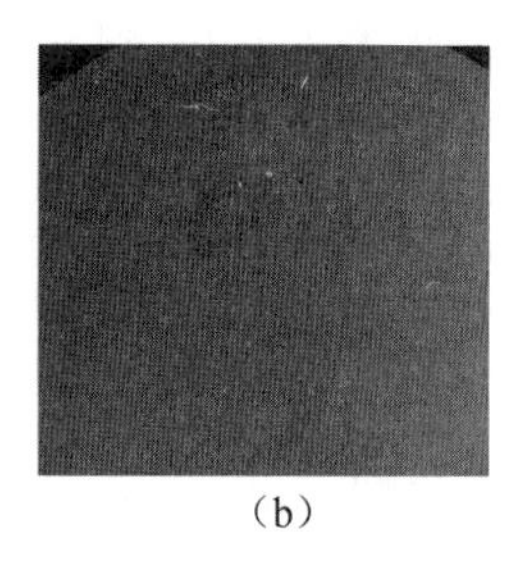
（b）

图 5－10　在日光和紫外线照射下，用 YNCDs 作为荧光油墨在滤纸上进行图像处理

（a）

（b）

图 5－11　用 YNCDs 作为荧光油墨在滤纸上书写后紫外线照射下的图像

5.4 小结

综上所述，本章研究了一种制备黄色发射氮掺杂 CDs(YNCDs) 的简便且低成本的方法。姜黄素可快速选择性淬灭 YNCDs 的荧光发射。在此基础上，构建了灵敏有效的姜黄素痕量检测荧光传感平台，其检出限为 0.01 $\mu mol \cdot L^{-1}$。同时，对荧光淬火机理进行了深入的分析和总结。研制的荧光探针在咖喱粉、人血清和尿液样品中均具有可行性，在医药和食品研究领域具有广阔的应用前景。考虑到优异的荧光特性，YNCDs 还可以作为信息加密的墨水。

参考文献

[1] Li T, E S, Wang J, et al. Regulating the properties of carbon dots via a solvent - involved molecule fusion strategy for improved sensing selectivity [J]. Analytica Chimica Acta, 2019, 1088: 107 - 115.

[2] Song W, Duan W, Liu Y, et al. Ratiometric detection of intracellular lysine and pH with one - pot synthesized dual emissive carbon dots [J]. Analytical Chemistry 2017, 89 (24): 13626 - 13633.

[3] Kou E, Yao Y, Yang X, et al. Regulation mechanisms of carbon dots in the development of lettuce and tomato [J]. ACS Sustainable Chemistry & Engineering, 2021, 9 (2): 944 - 953.

[4] Hola K, Sudolska M, Kalytchuk S, et al. Graphitic nitrogen triggers red fluorescence in carbon dots [J]. ACS Nano, 2017, 11 (12): 12402 - 12410.

[5] Ye X, Xiang Y, Wang Q, et al. A red emissive two - photon fluorescence probe based on carbon dots for intracellular pH detection [J]. Small, 2019, 15 (48): 1901673.

[6] Li T, Shi W, Mao Q, et al. Regulating the photoluminescence of carbon dots via a green fluorine - doping derived surface - state - controlling strategy [J]. Journal of Materials Chemistry C, 2021, 9: 17357 - 17364.

[7] Zhang W, Zhong H, Zhao P, et al. Carbon quantum dot fluorescent probes for food safety detection: Progress, opportunities and challenges [J]. Food Control, 2022, 133: 108591.

[8] Phukan K, Sarma R R, Dash S, et al. Carbon dot based nucleus targeted fluorescence imaging and detection of nuclear hydrogen peroxide in living cells [J]. Nanoscale Advances, 2022.

[9] Dong Y, Li T, Bateer B, et al. Preparation of yellow emissive nitrogen – doped carbon dots from o – phenylenediamine and their application in curcumin sensing [J]. New Journal of Chemistry, 2022, 46 (20): 9543 – 9549.

[10] Ge J, Jia Q, Liu W, et al. Red – emissive carbon dots for fluorescent, photoacoustic, and thermal theranostics in living mice [J]. Advanced Materials, 2015, 27 (28): 4169 – 4177.

[11] Liu J, Geng Y, Li D, et al. Deep red emissive carbonized polymer dots with unprecedented narrow full width at half maximum [J]. Advanced Materials, 2020, 32 (17): 1906641.

[12] Li T, Shi W, E S, et al. Green preparation of carbon dots with different surface states simultaneously at room temperature and their sensing applications [J]. Journal of Colloid and Interface Science, 2021, 591: 334 – 342.

[13] Hu Y, Gao Z. Sewage sludge in microwave oven: A sustainable synthetic approach toward carbon dots for fluorescent sensing of para – nitrophenol [J]. Journal of Hazardous Materials, 2020, 382: 121048.

[14] Liu J, Li D, Zhang K, et al. One – step hydrothermal synthesis of nitrogen – doped conjugated carbonized polymer dots with 31% efficient red emission for in vivo imaging [J]. Small, 2018, 14 (15): 1703919.

[15] Sun Z, Zhou W, Luo J, et al. High – efficient and pH – sensitive orange luminescence from silicon – doped carbon dots for information encryption and bio – imaging [J]. Journal of Colloid and Interface Science, 2022, 607: 16 – 23.

[16] Jiao Y, Gao Y, Meng Y, et al. One – step synthesis of label – free ratiometric fluorescence carbon dots for the detection of silver ions and glutathione and cellular imaging applications [J]. ACS Applied Materials & Interfaces, 2019, 11 (18): 16822 – 16829.

[17] Ding H, Yu S – B, Wei J – S, et al. Full – color light – emitting carbon dots with a surface – state – controlled luminescence mechanism [J]. ACS Nano, 2016, 10 (1): 484 – 491.

[18] Wang Q, Zhang S, Zhong Y, et al. Preparation of yellow – green – emissive carbon dots

and their application in constructing a fluorescent turn - on nanoprobe for imaging of selenol in living cells [J]. Analytical Chemistry, 2017, 89 (3): 1734 - 1741.

[19] Wang J, Li Q, Zheng J, et al. N, B - codoping induces high - efficiency solid - state fluorescence and dual emission of yellow/orange carbon dots [J]. ACS Sustainable Chemistry & Engineering, 2021, 9 (5): 2224 - 2236.

[20] Shan X, Chai L, Ma J, et al. B - doped carbon quantum dots as a sensitive fluorescence probe for hydrogen peroxide and glucose detection [J]. Analyst, 2014, 139 (10): 2322 - 2325.

[21] Li F, Li Y, Yang X, et al. Highly fluorescent chiral N - S - doped carbon dots from cysteine: Affecting cellular energy metabolism [J]. Angewandte Chemie International Edition, 2018, 57 (9): 2377 - 2382.

[22] Liu Q, Ren B, Xie K, et al. Nitrogen - doped carbon dots for sensitive detection of ferric ions and monohydrogen phosphate by the naked eye and imaging in living cells [J]. Nanoscale Advances, 2021, 3 (3): 805 - 811.

[23] Zhao J, Li F, Zhang S, et al. Preparation of N - doped yellow carbon dots and N, P co - doped red carbon dots for bioimaging and photodynamic therapy of tumors [J]. New Journal of Chemistry, 2019, 43 (16): 6332 - 6342.

[24] Zhou J, Shan X, Ma J, et al. Facile synthesis of P - doped carbon quantum dots with highly efficient photoluminescence [J]. RSC Advances, 2014, 4 (11): 5465 - 5468.

[25] Sarkar S, Das K, Ghosh M, et al. Amino acid functionalized blue and phosphorous - doped green fluorescent carbon dots as bioimaging probe [J]. RSC Advances, 2015, 5 (81): 65913 - 65921.

[26] Prasad K S, Pallela R, Kim D - M, et al. Microwave - assisted one - pot synthesis of metal - free nitrogen and phosphorus dual - doped nanocarbon for electrocatalysis and cell imaging [J]. Particle & Particle Systems Characterization, 2013, 30 (6): 557 - 564.

[27] 王海英，张宇琪，孙昊天，等. 姜黄素及其衍生物的作用及机制 [J]. 生理科学进展，2022，53 (4): 271 - 275.

[28] Bu L, Luo T, Peng H, et al. One - step synthesis of N - doped carbon dots, and their applications in curcumin sensing, fluorescent inks, and super - resolution nanoscopy [J]. Microchimica Acta, 2019, 186 (10): 675.

[29] Bian W, Wang X, Wang Y, et al. Boron and nitrogen co – doped carbon dots as a sensitive fluorescent probe for the detection of curcumin [J]. Luminescence, 2018, 33 (1): 174 – 180.

[30] Yang R, Mu W – Y, Chen Q – Y. Urazole – Au nanocluster as a novel fluorescence probe for curcumin determination and mitochondria imaging [J]. Food Analytical Methods, 2019, 12 (8): 1805 – 1812.

[31] Yang H, Li X, Wang X, et al. Silver – doped graphite carbon nitride nanosheets as fluorescent probe for the detection of curcumin [J]. Luminescence, 2018, 33 (6): 1062 – 1069.

[32] Du X, Wen G, Li Z, et al. Paper sensor of curcumin by fluorescence resonance energy transfer on nitrogen – doped carbon quantum dot [J]. Spectrochim Acta A Mol Biomol Spectrosc, 2020, 227: 117538.

[33] Yu C, Zhuang Q, Cui H, et al. A fluorescent "turn – off" probe for the determination of curcumin using upconvert luminescent carbon dots [J]. Journal of Fluorescence, 2020, 30 (6): 1469 – 1476.

[34] Liu L, Hua R, Zhang X, et al. Spectral identification and detection of curcumin based on lanthanide upconversion nanoparticles [J]. Applied Surface Science, 2020, 525.

[35] Anand S R, Bhati A, Saini D, et al. Antibacterial nitrogen – doped carbon dots as a reversible "fluorescent nanoswitch" and fluorescent Ink [J]. ACS Omega, 2019, 4 (1): 1581 – 1591.

[36] Korah B K, Murali A, Chacko A R, et al. Bio – inspired novel carbon dots as fluorescence and electrochemical – based sensors and fluorescent ink [J]. Biomass Conversion and Biorefinery, 2022.

[37] Li Y – X, Lee J – Y, Lee H, et al. Highly fluorescent nitrogen – doped carbon dots for selective and sensitive detection of Hg^{2+} and ClO^- ions and fluorescent ink [J]. Journal of Photochemistry and Photobiology A: Chemistry, 2021, 405: 112931.

[38] Atchudan R, Jebakumar Immanuel Edison T N, Perumal S, et al. Indian gooseberry – derived tunable fluorescent carbon dots as a promise for in vitro/in vivo multicolor bioimaging and fluorescent ink [J]. ACS Omega, 2018, 3 (12): 17590 – 17601.

[39] Siddique A B, Singh V P, Pramanick A K, et al. Amorphous carbon dot and chitosan based composites as fluorescent inks and luminescent films [J]. Materials Chemistry

and Physics, 2020, 249: 122984.

[40] Nandi N, Sarkar P, Sahu K. N – doped carbon dots for visual recognition of 4 – nitroaniline and use in fluorescent inks [J]. ACS Applied Nano Materials, 2021, 4 (9): 9616 – 9624.

[41] He C, Xu P, Zhang X, et al. The synthetic strategies, photoluminescence mechanisms and promising applications of carbon dots: Current state and future perspective [J]. Carbon, 2022, 186: 91 – 127.

[42] Liu X, Li T, Hou Y, et al. Microwave synthesis of carbon dots with multi – response using denatured proteins as carbon source [J]. RSC Advances, 2016, 6 (14): 11711 – 11718.

[43] Liu X, Li T, Wu Q, et al. Carbon nanodots as a fluorescence sensor for rapid and sensitive detection of Cr(VI) and their multifunctional applications [J]. Talanta, 2017, 165: 216 – 222.

[44] Dong Y, Li J, Zhang L. 3D hierarchical hollow microrod via in – situ growth 2D SnS nanoplates on MOF derived Co, N co – doped carbon rod for electrochemical sensing [J]. Sensors and Actuators B: Chemical, 2020, 303: 127208.

[45] Sun S, Zhang L, Jiang K, et al. Toward high – efficient red emissive carbon dots: Facile preparation, unique properties, and applications as multifunctional theranostic agents [J]. Chemistry of Materials, 2016, 28 (23): 8659 – 8668.

[46] Miao X, Yan X, Qu D, et al. Red emissive sulfur, nitrogen codoped carbon dots and their application in ion detection and theraonostics [J]. ACS Applied Materials & Interfaces, 2017, 9: 18549? 18556.

[47] Pan D, Zhang J, Li Z, et al. Hydrothermal route for cutting graphene sheets into blue – luminescent graphene quantum dots [J]. Advanced Material, 2010, 22 (6): 734 – 738.

[48] Li Y, Hu Y, Zhao Y, et al. An electrochemical avenue to green – luminescent graphene quantum dots as potential electron – acceptors for photovoltaics [J]. Advanced Materials, 2011, 23 (6): 776 – 780.

[49] Liu R, Wu D, Feng X, et al. Bottom – up fabrication of photoluminescent graphene quantum dots with uniform morphology [J]. Journal Of the American Chemical Society, 2011, 133 (39): 15221 – 15223.

[50] Dong Y, Pang H, Yang H B, et al. Carbon – based dots co – doped with nitrogen and

sulfur for high quantum yield and excitation – independent emission [J]. Angewandte Chemie International Edition, 2013, 52 (30): 7800 –7804.

[51] Meng Y, Jiao Y, Zhang Y, et al. One – step synthesis of red emission multifunctional carbon dots for label – free detection of berberine and curcumin and cell imaging [J]. Spectrochimica Acta Part A: Molecular and Biomolecular Spectroscopy, 2021, 251: 119432.

[52] Ma H, Sun C, Xue G, et al. Facile synthesis of fluorescent carbon dots from Prunus cerasifera fruits for fluorescent ink, Fe^{3+} ion detection and cell imaging [J]. Spectrochimica Acta Part A: Molecular and Biomolecular Spectroscopy, 2019, 213: 281 –287.

[53] Zhou X, Zhao G, Tan X, et al. Nitrogen – doped carbon dots with high quantum yield for colorimetric and fluorometric detection of ferric ions and in a fluorescent ink [J]. Microchimica Acta, 2019, 186 (2): 67.

[54] Bandi R, Devulapalli N P, Dadigala R, et al. Facile conversion of toxic cigarette butts to N,S – Codoped carbon dots and their application in fluorescent film, security Ink, bioimaging, sensing and logic gate operation [J]. ACS Omega, 2018, 3 (10): 13454 –13466.

[55] Paul S, Gayen K, Nandi N, et al. Carbon nanodot – induced gelation of a histidine – based amphiphile: application as a fluorescent ink, and modulation of gel stiffness [J]. Chemical Communications, 2018, 54 (34): 4341 –4344.

第6章
水热法制备黄光氮掺杂碳点及其应用

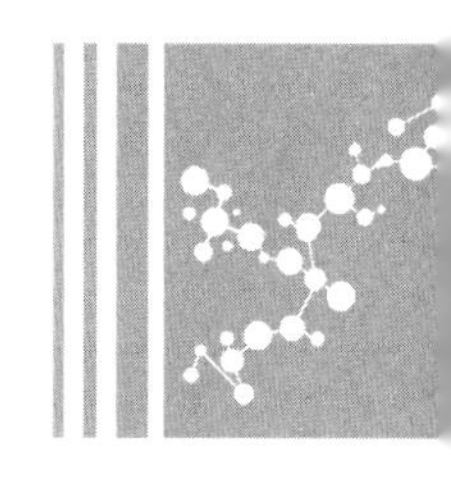

6.1 引言

当前全球水资源短缺,可用的淡水不足地球上全部水资源的0.5%,据报道全球用水量是人类人口增长率的2倍多[1-4]。水是生命的源泉,是生产要素。水资源体系的健康直接影响到社会的可持续发展[5]。在过去的一段时间,中国的水资源体系推动了社会经济的快速发展[6]。随着城市工业化的进步,水污染问题日益突出,这导致环境问题已经成为世界上的一个大问题,其中严重的环境问题是存在于工业工厂中的染料废水。我国作为水资源严重短缺大国,随着纺织工业的发展以及印刷废水排放量逐年增加,我国水资源的短缺问题日益严重。由于合成染料的耐用性是为了满足消费市场的需求而设计的,导致染料能够抵抗生物降解。染料种类复杂多样,根据结构不同可分芳甲烷染料、苯酚染料、蒽醌染料、亚硝基染料以及偶氮染料等。三苯甲烷类染料作为第三大类染料,其因具有价格低廉、色度高、染色效果好等优点而被广泛应用。其中,结晶紫(CV)是三苯甲烷类染料中应用最广泛的一种染料,在应用方面主要涉及药学、纺织、军事等领域。结晶紫是一种含有芳香环的蛋白质染料,广泛应用于纺织、造纸和制药行业[7-9]。此前,它也作为杀菌剂应用于水产养殖[10]。但是,研究发现CV对生物体有毒性并可能引起人体癌变,因此目前禁止在水产养殖中使用[11]。由于其有害的影响,对CV的敏感和选择性检测的需求很高。在过去的几年里,人们已经开发出一些测定CV的方法,包括:免疫吸附法、微萃取耦合高效液相色谱法、液相色谱耦合电喷雾电离串联质谱法等多种方法[12-18]。然而,这些方法存在仪器复杂、预处理步骤烦琐、检测时间长等问题。荧光法具有简单、快速的优点,例如:基于荧光$YVO_4:Eu^{3+}$和二氧化硅纳米颗粒实现了CV检测[14,19,20]。然而,$YVO_4:Eu^{3+}$的成本高,二氧化硅纳米粒子的量子产率低(0.43%)极大限制了它的应用。因此,设计制备廉价且高效的荧光探针对CV的精确分析是十分必要的。

CDs是一种横向尺寸低于20 nm的新兴纳米碳材料,由于其独特的光学、化学和物理特性,有望促进传感、成像、生物医疗、催化和能量转换等领域的发展[13,21-26]。CDs最迷人的特征之一是荧光。然而,大多数CDs发射的光波长较短,通常位于蓝到绿区域[27,28]。荧光发射峰超过540 nm的黄色或红色发射CDs难以获得,这极大地阻碍了CDs的多领域应用,特别是在生物成像、照明设备和传感等领域,因此对制备出长波长发射CDs的需求非常迫切。通过选择合适的前驱体或修改表面态,研究人员将CDs的发射扩展到长波长[29-31]。例如,Wang等人报道了硫掺杂的黄色发光CDs[32]。Lin等人通过溶剂热处理1,2,4－三氨基苯合成了黄色发射CDs[31]。尽管取得了一些令人欣喜的进步,但具有高量子产率的长波长发射CDs的快速制备仍处于早期阶段,需要进行大量的工作。

CDs由于其卓越的光学性质、化学稳定性和水溶性受到了化学检测领域的广泛关注。目前,CDs被大量应用于化学检测之中,如对金属离子、阴离子、有机小分子等进行定量检测[33]。例如,Zhang等以B族维生素作为碳源和氮源制备了一种含氮量丰富的高量子产率碳点[34]。研究发现,这个碳量子点可以与Hg^{2+}进行特异性结合,从而实现对Hg^{2+}检测。实验结果显示,其检出限为0.23 nmol·L^{-1}。除此之外,科研人员还开发了碳点基探针用于Cu^{2+}、Fe^{3+}、Pb^{2+}等许多金属离子检测的方法[35-37]。随着CDs基探针开发对金属离子检测的蓬勃发展,人们也开始聚焦用碳点对阴离子和小分子进行检测。例如,Zong等利用球形介孔二氧化硅作为纳米反应器,通过添加柠檬酸和三种无机离子(NaCl、LiCl和KNO_3)超声制备了一种碳点[38]。此碳点能够与Cu^{2+}特异性结合,这会导致碳量子点发生荧光猝灭。然而,当向碳点－铜体系中加入半胱氨酸时,Cu^{2+}将从碳点表面释放,从而荧光恢复。最终,实现了此碳点既能用于检测Cu^{2+},又能用于检测半胱氨酸。最近,Hu等人报道了一种新型黄色发光的硒氮掺杂碳点对CV的快速检测,在他们的研究中通过水热法,使用硒脲和o－PD作为前体制备了在566 nm具有强光发的CDs,研究发现制备CDs的黄色荧光可以被CV选择性猝灭,这使制备的CDs有望用于CV传感[13]。Zhao报道了仅使用1,2,4,5－四氨基苯为碳源和氮源的一步溶剂热法制备多色荧光CDs,所制备的CDs的发射波长在527~605 nm的范围内,量子产率达到10.0%~47.6%,并成功地用作荧光油墨[39]。通过多种表征方法比较了制备的红光CDs(R－CDs,λ_{em} = 605 nm)和黄光CDs(Y－CDs,λ_{em} = 543 nm),并研究了它们的发光机理[39]。研究发现粒径大小、石墨氮和含氧官能团有利于长波长发射CDs的形成。其中,Y－CDs对结晶紫具有响应性,其荧光可以被结晶紫猝灭。因此,利用该现象来开发了线性范围为0.1~11 μmol·L^{-1},检测限为20 nmol·L^{-1}的结晶紫的检测方法。上述证明了制备长波长发射CDs可以有效地用于CV检测中。

水热法是一种简单、快捷的制备CDs方法。该方法所用碳源广泛,从有机化合物到生活废弃物都可以作为碳源。因此,水热法也是目前制备CDs最常用的方法之一。水热法是将前驱体的溶液密封在反应釜中,通过高温高压反应制得CDs。本研究中采用水热法,以o－PD为原料,制备了强黄光CDs(SYCDs)。结晶紫作为一种危险染料,对环境和人类健康构成严重威胁。因此,本章开发了一种基于SYCDs为探针对结晶紫进行灵敏的检测方法。研究发现,结晶紫可以有效抑制SYCDs的荧光发射。因此,以SY-

CDs 为纳米探针，建立了检测结晶紫的传感器平台，在 0.02 ~ 15 μmol · L^{-1} 范围内线性良好其检出限(3S/N)为 0.006 μmol · L^{-1}。通过详细的研究，提出了内滤效应是 SYCDs 对结晶紫的传感机理。进一步论证了基于 SYCDs 的鱼组织样品 CV 定量评价方法的可行性。

6.2　实验部分

6.2.1　实验仪器

KQ - 100B/800KDE 型超声波清洗器(中国昆山市超声仪器有限公司)；
BSA22AS 单盘型分析电子天平(中国北京赛多利斯仪器有限公司)；
TG16 - WS 台式高速离心机(中国湘仪实验室仪器开发有限公司)；
2XZ - 2 型真空泵(中国临海市谭式真空设备有限公司)；
PB - 10 标准型 pH 计(中国北京赛多利斯仪器有限公司)；
Lambda Bio20 紫外可见光分光光度计(美国珀金埃尔默仪器有限公司)；
F - 7000 荧光分光光度计(日本日立公司)；
Bruker Dimension icon 原子力显微镜(德国布鲁克公司)；
DZF - 6020 型真空干燥箱(中国上海精宏实验设备有限公司)；
One Spectra 红外光谱仪(美国珀金埃尔默仪器有限公司)；
DHG - 9037A 电热恒温干燥箱(中国上海精宏实验设备有限公司)；
YE5A44 型手动可调式移液器(中国上海大龙医疗设备有限公司)；
EscaLab 250Xi X 射线光电子能谱分析仪(美国赛默飞世尔公司)；
Malvern Nano - ZS 粒度仪(英国马尔文公司)；

6.2.2　实验试剂

本章所使用化学试剂和品牌如下：

O - PD、Cys、葡萄糖、抗坏血酸(AA)、谷胱甘肽(GSH)以及本章使用的其他氨基酸购自阿拉丁化学有限公司(中国上海)。本章使用的无机盐购自国药集团化学试剂有限公司(中国上海)。

除特别声明外，所有试剂皆为分析纯且未经任何前处理。实验用水为二次去离子水(18 MΩ cm)。

6.2.3 SYCDs 的制备方法

将0.2 g o-PD 加入到20 mL 去离子水中,超声10 min后,将溶液转移到50 mL 的聚四氟乙烯内胆高压釜中,在140 ℃加热12 h。冷却至室温后,将所得溶液通过0.22 μm过滤膜去除大颗粒。之后,在75 ℃旋转蒸发去除溶剂得到粗产物。然后,将产物溶于去离子水中,通过纤维素酯膜(MWCO 相对分子质量:500~1 000)透析48 h除去未反应的物质。然后收集溶液并冷冻干燥获得 CDs,将其命名为 SYCDs。

6.2.4 SYCDs 的表征方法

本章通过 U-3900 紫外可见分光光度计使用1 cm 光程的比色皿,在扫描间隔为1 nm条件下测得紫外-可见(UV-vis)吸收光谱。荧光光谱是通过 F-7000 荧光光谱仪使用1 cm 光程的比色皿测得,激发和发射狭缝均设置为10 nm。傅里叶红外(FT-IR)光谱使用 Nicolet-6700 红外光谱仪采用溴化钾压片法测定并记录1 000~4 000 cm^{-1}的数据。X 射线光电子能谱(XPS)在配备 Al Kα 280.00 eV 激发光源的 ESCALAB 250 表面分析系统上进行测得。

6.2.5 SYCDs 的稳定性测试

光稳定性测试:取一定量的 SYCDs 溶液,在365 nm 波长光下照射0、1、2、3、4、5、6小时。在 SYCDs 最佳激发波长下,测试上述溶液荧光发射光谱。每组样品测试3次。

离子强度稳定性测试:取6组 SYCDs 溶液,分别加入不同浓度的 NaCl,使 SYCDs 溶液中 Na^+浓度分别为:0、0.2、0.4、0.6、0.8、1.0 mol·L^{-1}。在 SYCDs 最佳激发波长下,测试上述溶液荧光发射光谱。每组样品测试3次。

6.2.6 SYCDs 用于结晶紫定量检测

通过以下步骤用 SYCDs 作为荧光探针进行结晶紫测定。首先制备结晶紫原液,然后用磷酸缓冲溶液(pH=7.0)稀释10倍。将1.0 mL 不同浓度的结晶紫溶液与1.0 mL SYCDs 溶液(20.0 μg/mL,在磷酸缓冲溶液中制备)混合。之后,在410 nm 的激发下测试荧光发射光谱。以结晶紫浓度为横坐标,荧光淬灭率$(F_0-F)/F_0$为纵坐标绘图,其中F_0和 F 分别代表添加结晶紫前后 SYCDs 的荧光强度。此外,在相同条件下将结晶紫替换为干扰物研究了 SYCDs 的选择性。所有测试均重复3次。

本研究以 SYCDs 作为探针通过加标回收法对鱼中的结晶紫进行了定量分析。鱼组织在使用前通过文献报道的方法进行预处理[20]。鱼的肌肉被切碎冷冻。之后,取10 g解冻肌肉放入30 mL乙腈中,超声60分钟,收集上清液并离心。然后,在混合物中加入无水硫酸钠。将蒸发乙腈后的样品分散于磷酸盐缓冲液中,而后将样品与含有不同浓度的CV溶液和的SYCDs分散液混合,进行CV分析。

6.3 结果与讨论

6.3.1 SYCDs的制备与表征

本研究以o-PD为碳源,水为溶剂,无须使用特殊设备,无须添加额外氧化剂,仅利用高温下o-PD的自聚合制备SYCDs。o-PD上的$-NH_2$基团具有较高的活性,使o-PD在高温高压下容易发生氧化或聚合反应,生成o-PD二聚体、o-PD三聚体和2,3-二硝基苯胺等[25]。随着反应的持续进行,o-PD的聚合反应不断形成长链聚o-PD,这些聚合物可以纠缠在一起,最终形成SYCDs。

与之前有关碳点的报道一致,本文通过FT-IR、XPS、粒径测试对SYCDs结构进行了表征。在分子中,组成化学键或官能团的原子处于不断振动的状态,其振动频率与红外光的振动频率相当。当用红外光照射分子时,分子中的化学键或官能团可发生振动吸收,不同的化学键或官能团吸收频率不同,在FT-IR上将处于不同位置,从而可获得分子中含有何种化学键或官能团的信息。图6-1为FT-IR图展示了SYCDs的主要官能团,包括$-OH/-NH_2$的伸缩振动(3 000~3 750 cm^{-1}),C-H的伸缩振动(约2 900和2 850 cm^{-1}),C=C的伸缩振动(约1 650 cm^{-1}),C-H的弯曲振动(约1 390 cm^{-1})以及C-O的伸缩振动(约1 200和1 050 cm^{-1})这些说明了SYCDs中羟基、羰基和氨基的存在,同时上述结果也表明SYCDs结构中聚芳族结构的存在[25,27,28,40]。

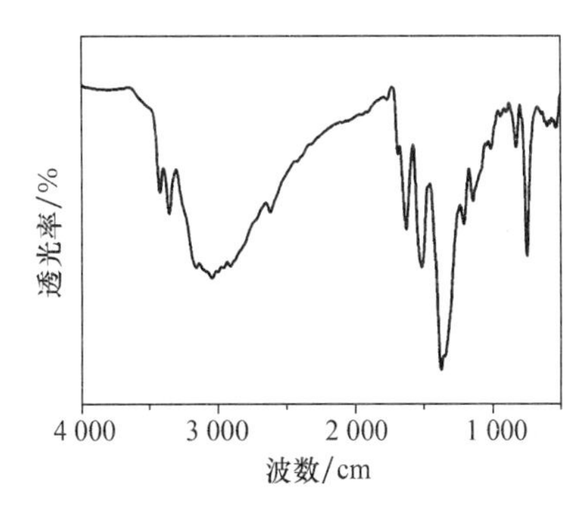

图6-1 SYCDs的傅里叶红外光谱

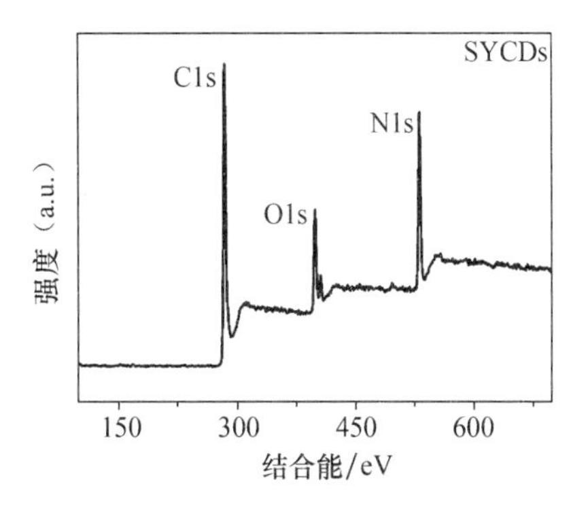

图6-2 SYCDs的XPS谱

XPS 的原理是用 X 射线去辐射样品,使原子或分子的芯电子或价电子受激发射出来。被光子激发出来的电子称为光电子。可以测量光电子的能量,以光电子的动能/束缚能为横坐标,相对强度为纵坐标可做出光电子能谱图,从而获得试样有关信息。XPS 因对化学分析最有用,因此被称为化学分析用电子能谱。本章为了深入认识 SYCDs 上所具有的官能团,采用 XPS 对 SYCDs 进行表征。如图 6-2,图中在 289.0、400.0 和 532.0 eV 处存在 3 个峰,这分别归属于 C 1s、N 1s、O 1s。这个结果证明在制备的 SYCDs 中包含有碳、氮、氧 3 种元素。

在高分辨率 XPS 光谱中,C 1s(图 6-3)可以被分成为 5 个结合能峰,包括:284.0 eV时的 C=C/C-C、285.1 eV 时的 C-N、286.0 eV 时的 C-O、287.8 eV 时的 C=N 和 289.0 eV 时的 CO-NH[41]。N 1s 谱图(图 6-4)显示在 398.5 eV、399.2 eV、400.1 eV 和401.1 eV处有 4 个峰,分别属于吡啶氮、氨基氮、吡咯氮和石墨氮[42]。O1s 谱图(图 6-5)分别在 531.9 和 532.8 eV 处包含 2 个峰对应 C=O 和 C-O[42]。上述 XPS 结果与FT-IR 光谱结果一致,进一步证明了 SYCDs 中这些官能团的存在,表明 SYCDs 成功制备。

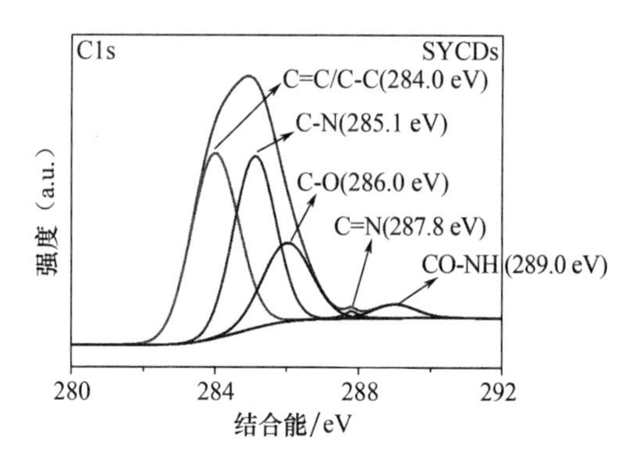

图 6-3　SYCDs 的 C 1s 的 XPS 高分辨率谱

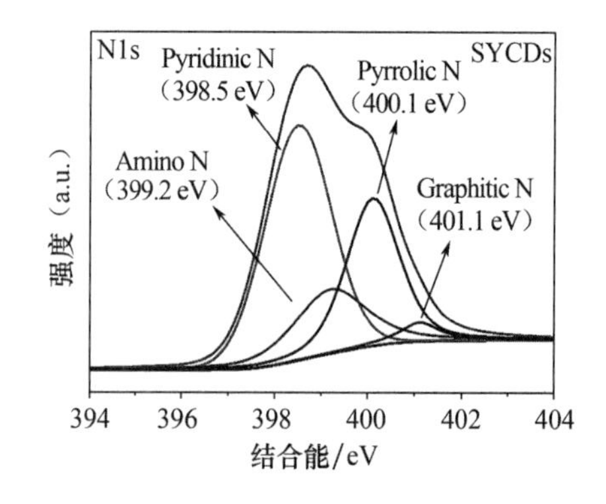

图 6-4　SYCDs 的 N 1s 的 XPS 高分辨率谱

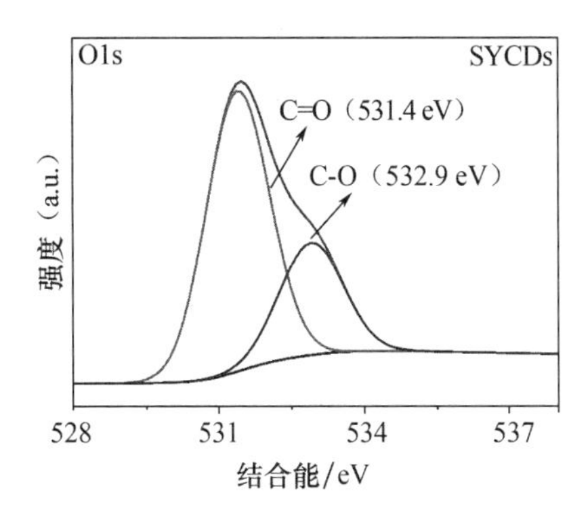

图6-5　SYCDs的O 1s的XPS高分辨率谱

CDs被认为是一种尺寸低于20 nm的准零维纳米碳材料,为了认识SYCDs的粒径大小,图6-6展示了SYCDs的粒径分布图,从图中可以看到SYCDs的粒径范围为2~7 nm,平均粒径约4.8 nm。上述结果证实SYCDs的尺寸与文献所报道的大多数CDs尺寸一致[43-46]。

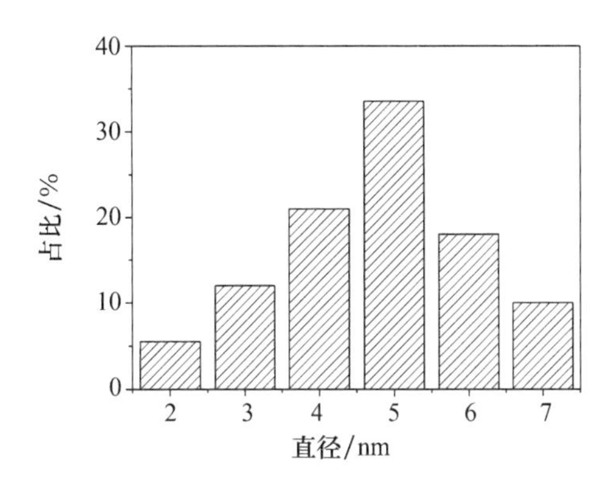

图6-6　SYCDs的粒径分布图

6.3.2　SYCDs的光学性质

在化合物分子中存在单键的σ电子、有形成双键的π电子和有未成键的孤对n电子。当分子吸收一定能量的辐射能时,这些电子就会跃迁到较高的能级,此时电子所占的轨道称为反键轨道,而这种电子跃迁同内部的结构有密切的关系,这就产生了UV-Vis光谱。从图6-7中可以看到,SYCDs的UV-Vis光谱图展现了碳点的典型UV区域吸收一直延长到可见光区域,这与相关文献报道中碳点的UV-Vis光谱图一致[19]。从图中还可以看到2个主要的吸收峰:SYCDs在200~360 nm高能区中均表现出了C=

C 引起的 $\pi-\pi^*$ 跃迁和 C = O/C = N 引起的 $n-\pi^*$ 跃迁的吸收带。除了这些吸收带，在 360 ~ 700 nm 的低能区也观察到了吸收带，这归因于共轭 C = N 的 $n-\pi^*$ 跃迁[47,48]。

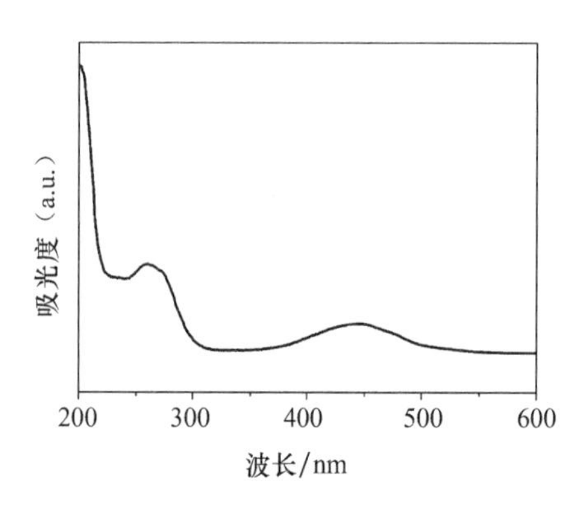

图 6 - 7　SYCDs 的紫外 - 可见光光谱

碳点的荧光性能无论对其在基础研究和实际应用中都是一个重要的性质。如图 6 - 8所示，通过一系列递增激发光照射 SYCDs 溶液，得到了 SYCDs 溶液的荧光发射光谱。图中显示了 SYCDs 存在单个发射中心，发射最大值在 555 nm 处，随着发射波长从 380 到 450 nm 的变化，发光强度先增大后减小，在发射波长 410 nm 时达到最大值。但与传统 CDs 相比，SYCDs 展现了不同的发射特性，即发射位置不随激发波长的变化而发生蓝/红位移[49 - 51]。产生这个结果通常归因于表面态发射[41,52]。

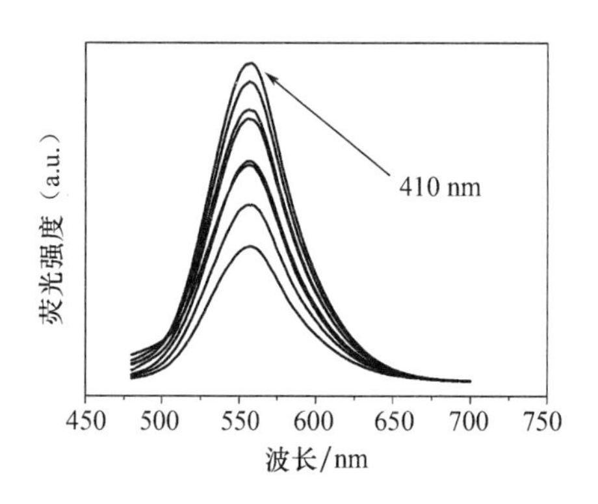

图 6 - 8　不同激发波长下 SYCDs 的荧光发射光谱

如上一章节所述，对于荧光材料来说，荧光稳定性是非常重要的。在一些极端条件下，如高离子强度或者是长期光照，这些都会使一些荧光材料发生荧光淬灭。荧光淬灭的发生不但影响了荧光材料的荧光性，而且更多的影响了荧光材料的应用。鉴于以上原因，对于荧光材料的稳定性测试显得尤为重要。SYCDs 在高离子强度条件下和光照下还表现出了极强的稳定性。如图 6 - 9 所示，实线部分为 SYCDs 在不同 NaCl 浓度下

($0.0 \sim 1.0\ mol \cdot L^{-1}$)归一化后的最大荧光发射强度。实际上,由特定碳源制备得到的碳量子点会保留了原始碳源的重要官能团[28]。在这个体系中,SYCDs是由o-PD制备而成的,因此在SYCDs上保留了一些苯环和氨基官能团。这些SYCDs上的官能团将在一定程度上降低小型电荷作用[28]。除此之外,在光照6 h之后,SYCDs的归一化后的最大荧光发射强度(图6-9虚线部分)高于0.9,这展现了SYCDs良好的反淬灭性质。

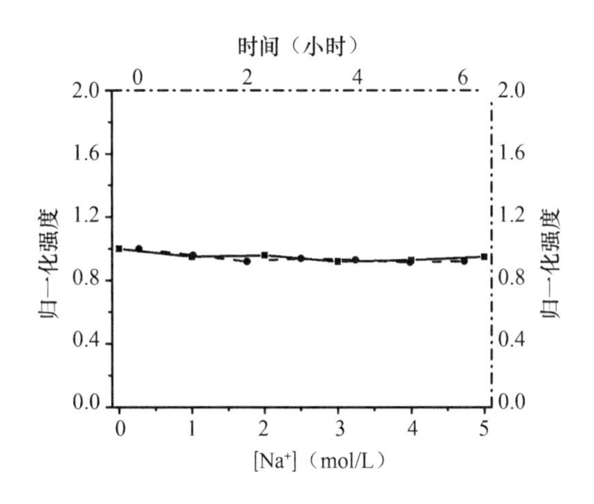

图6-9 在不同NaCl浓度(实线),持续的紫外光照射下(虚线)条件下SYCDs的最大荧光强度

6.3.3 基于SYCDs的结晶紫荧光检测

结晶紫是一种含有芳香环的蛋白质染料,广泛应用于纺织、造纸和制药行业[7,8]。此前,它也作为杀菌剂应用于水产养殖[10]。但是,研究发现CV对生物体有毒性并可能引起人体癌变。研究发现,CV可以有效地抑制SYCDs的荧光发射。因此,本章以SY-CDs为探针用于荧光检测CV。图6-10展示了添加一系列不同浓度CV后($0 \sim 50\ \mu mol \cdot L^{-1}$),SYCDs的荧光强度变化。从图6-11可以看出,SYCDs的荧光随着CV浓度的增加而降低。当CV浓度在$0.02 \sim 15\ \mu mol \cdot L^{-1}$的范围内,淬灭率$(F_0 - F)/F_0$与CV浓度之间具有良好的线性关系,线性方程为:$(F_0 - F)/F_0 = 0.034\,4C + 0.013\,8$,($R^2 = 0.993$)。当信噪比为3时,检出限为$0.006\ \mu mol \cdot L^{-1}$。与其他CV检测方法相比(表6-1),该传感体系具有操作简单、线性范围宽、灵敏度高等优点。

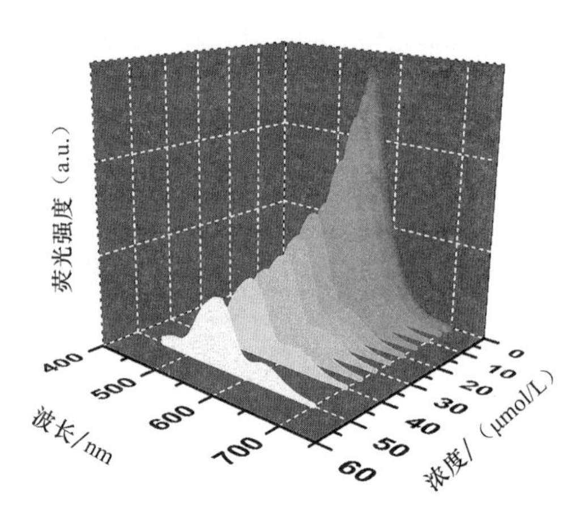

图 6-10　不同浓度（0 ~ 50.0 μmol · L^{-1}）结晶紫下 SYCDs 的荧光光谱

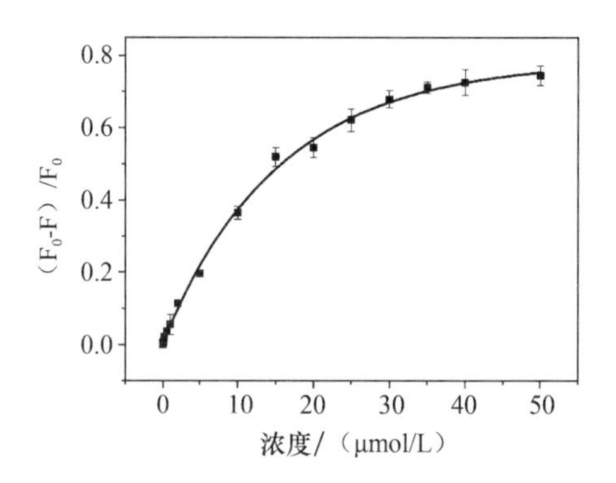

图 6-11　SYCDs 的（F_0 − F）/F_0 与结晶紫浓度的关系

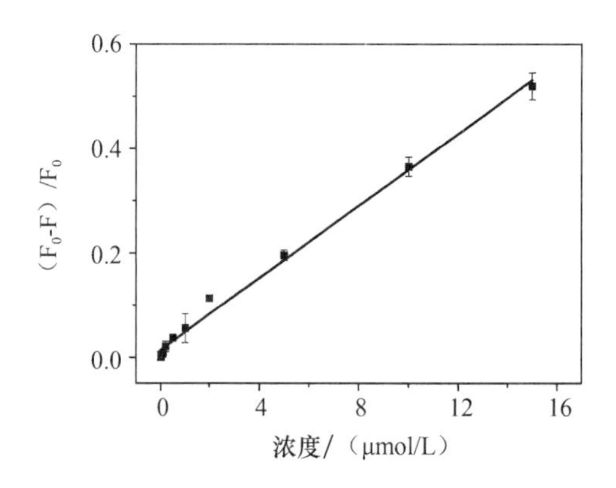

图 6-12　结晶紫浓度在 0.02 ~ 15.0 μmol · L^{-1} 范围内（F_0 − F）/F_0 与结晶紫浓度的线性相关图

本实验进一步研究了SYCDs对CV的选择性。如图6－13所示，Al^{3+}、Ca^{2+}、Co^{2+}、Ni^{2+}、Cu^{2+}、Cd^{2+}、Hg^{2+}、Zn^{2+}、Mn^{2+}、Pb^{2+}、Fe^{2+}、Fe^{3+}、Cr^{3+}等典型金属阳离子和NO^{3-}、Cl^-、PO_4^{3-}、SO_4^{2-}等常见阴离子浓度与CV浓度相同时(50 μmol·L^{-1})，对测定不产生干扰。一些重要的生理分子，包括葡萄糖(Glu)、抗坏血酸(AA)、尿酸(UA)、多巴胺(DA)、谷胱甘肽(GSH)和半胱氨酸(Cys)，在相同的条件下也不会引起SYCDs明显的荧光淬灭。这些发现证明了CDs对CV的特异性荧光反应。

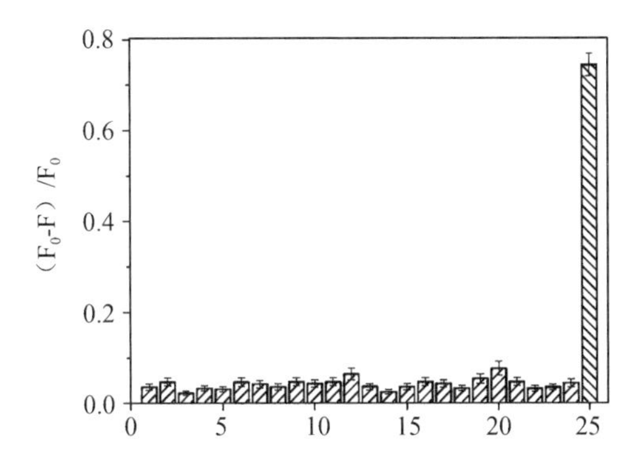

图6－13 利用SYCDs对结晶紫进行选择性检测

(从左至右分别为：Al^{3+}、Ca^{2+}、Co^{2+}、Ni^{2+}、Na^+、Cu^{2+}、Cd^{2+}、Hg^{2+}、Zn^{2+}、Mn^{2+}、Pb^{2+}、Fe^{2+}、Fe^{3+}、Cr^{3+}、NO^{3-}、Cl^-、PO_4^{3-}、SO_4^{2-}、葡萄糖(Glu)、抗坏血酸(AA)、尿酸(UA)、多巴胺(DA)、谷胱甘肽(GSH)、半胱氨酸(Cys)和CV)

为了测试SYCDs在实际样品中对CV的检测能力，将SYCDs应用于鱼组织中CV的分析。结果如表6－1所示。这些食用鱼类样本中未检出CV。通过加标回收法测定不同样品中CVI的浓度，分析结果在表6－2中。从表中可以看到实际样品回收率为98.1%～102.3%，RSD为1.2%～2.2%。这个结果说明SYCDs可以应用于实际检测中。

表6－1 对比一些已经报道的检测Cr(VI)的检测方法

探针	检出限/(μmol·L^{-1})	线性范围/(nmol·L^{-1})	参考文献
CDs	0.02～1.60	7.3	[13]
$Ru(bpy)_3^{2+}$/TPA	0.003～0.55	0.11	[53]
胶体银	0.01～0.8	3.6	[54]
Y－CDs	0.1～11	20	[39]

续表

探针	检出限/(μmol·L⁻¹)	线性范围/(nmol·L⁻¹)	参考文献
硅纳米颗粒	0.18～36.76	61.3	[20]
$YVO_4:Eu^{3+}$	0.3～12.0	83	[19]
多克隆抗体	0.005～0.490	3.3	[12]
SYCDs	0.02～15	6	本工作

表 6－2　鱼组织中 CV 分析

样品名称	加标量/(μmol·L⁻¹)	回收量/(μmol·L⁻¹)	相对标准偏差/%	回收率/%
鲤鱼	0	—	—	—
	1	0.981	1.3	98.1
	3	3.02	1.5	100.6
	5	4.95	2.2	99.0
	10	10.23	1.9	102.3

6.3.4　结晶紫检测的淬灭机理

通常来说,结晶紫对 CDs 的荧光淬灭过程与内滤效应(IFE)有关[13,55]。本章中为了证实姜黄素对 SYCDs 的淬灭机理是 IFE,对 SYCDs 和姜黄素进行了光学性能测试。如图 6－14 所示,SYCDs 的荧光发射光谱与结晶紫的吸收光谱几乎完全重叠,荧光共振能量转移(FRET)和 IFE 两种淬灭机理均依赖于荧光团的激发或发射区域与淬灭剂的吸收区域一定程度的重叠[13]。事实上,在 FRET 过程中激发态能量/电子转移的发生会显著改变荧光团的寿命[56]。在图 6－15 中,SYCDs 的荧光寿命约为 3.1 ns,姜黄素的加入几乎不会对 SYCDs 的荧光寿命(3.13 ns)造成任何改变。这个结果说明 FRET 并不是姜黄素对 SYCDs 的荧光淬灭的机理。

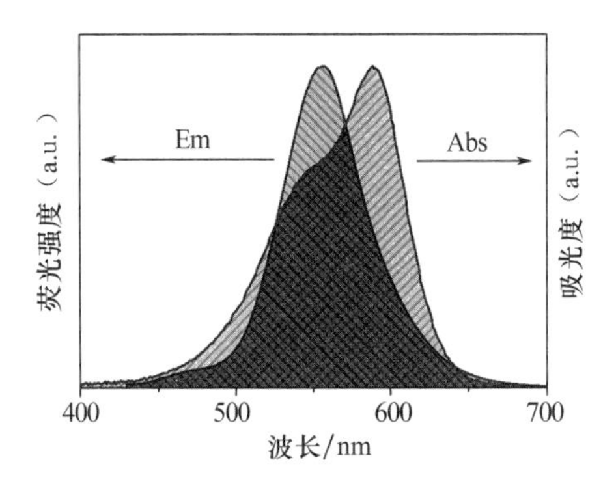

图6-14　结晶紫的吸收光谱和SYCDs的荧光发射光谱

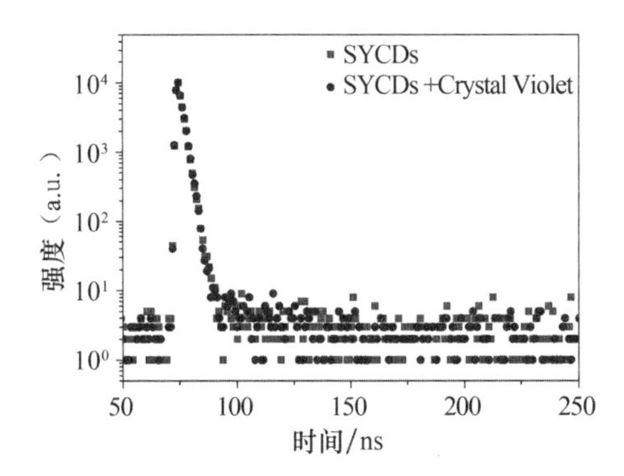

图6-15　结晶紫添加前后SYCDs的荧光寿命衰减曲线

图6-16显示，SYCDs对CV的荧光响应与温度无关，这证实了IFE在SYCDs淬灭过程中的作用[13,57]。另外，从图6-17可以发现，当加入足够的CV后，SYCDs荧光光谱的形状发生了变化。这是由于CV作为荧光吸收剂，在591 nm处吸收最大，650~750 nm吸收不足，因此SYCDs/CV混合物在640 nm左右出现了一个新的荧光峰，这种现象也可以归因于IFE[13,57]。

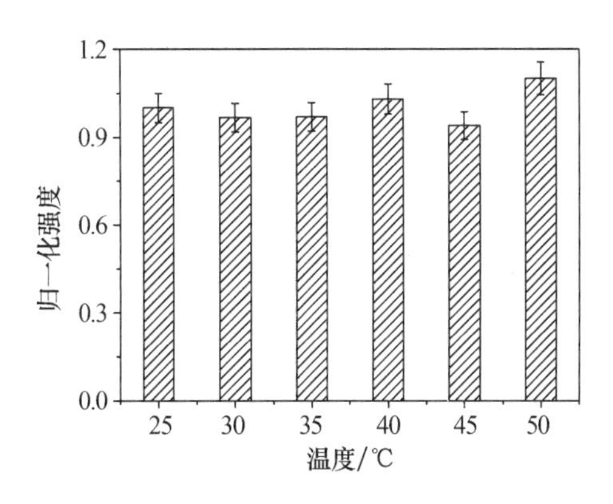

图6-16　不同温度下加入结晶紫后SYCDs的归一化荧光强度

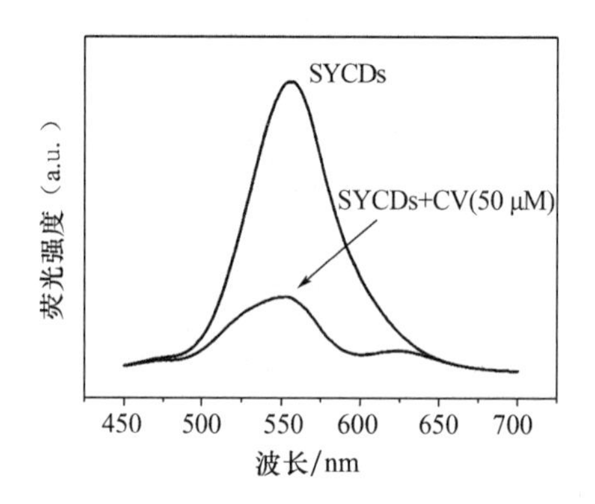

图 6－17　结晶紫添加前后 SYCDs 的荧光发射图

6.4　小结

本章 o－PD 为碳源，采用简单、快速、方便的微波法成功制备了 SYCDs。同时，对 SYCDs 进行了表征和性质研究。得出的结论如下：①SYCDs 被成功制备，并通过 UV－Vis、FT－IR、XPS 粒径测试对 SYCDs 所具有的官能团、形貌以及尺寸进行了表征；②制备的 SYCDs 展现了较强的荧光性，其 ILCDs 的最佳激发波长为 401 nm，最大发射峰位于 555 nm；③制备的 SYCDs 在高离子强度下展示了良好的稳定性，同时在长时光照下 SYCDs 荧光性不易被漂白，SYCDs 所展现的这些稳定性为其在实际应用中奠定了基础；④SYCDs 可以用于对 CV 选择性检测，其检测的线性范围是 0.02～15.00 $\mu mol \cdot L^{-1}$，检出限为 0.006 $\mu mol \cdot L^{-1}$。同时，在实际样品中用 SYCDs 检测 CV 的回收率为 98.1%～102.3%，RSD 为 1.2%～2.2% 之间，实验结果让人满意。这表明 ILCDs 可以应用于实际样品检测；⑤CV 对 SYCDs 的淬灭机理为 IFE。

参 考 文 献

[1] Han F, Kambala V S R, Srinivasan M, et al. Tailored titanium dioxide photocatalysts for the degradation of organic dyes in wastewater treatment: A review [J]. Applied Catalysis A: General, 2009, 359 (1): 25－40.

[2] Shevah Y. Water scarcity, water reuse, and environmental safety [J]. 2014, 86 (7): 1205－1214.

[3] Leal Filho W, Totin E, Franke J A, et al. Understanding responses to climate – related water scarcity in Africa [J]. Science of The Total Environment, 2022, 806: 150420.

[4] He C, Liu Z, Wu J, et al. Future global urban water scarcity and potential solutions [J]. Nature Communications, 2021, 12 (1): 4667.

[5] Garrote L, Sordo – Ward A. Preface to the special issue: Managing water resources for a sustainable future [J]. Water Resources Management, 2020, 34 (14): 4307 – 4311.

[6] Varis O, Kummu M, Lehr C, et al. China's stressed waters: Societal and environmental vulnerability in China's internal and transboundary river systems [J]. Applied Geography, 2014, 53: 105 – 116.

[7] Yakout S M, Hassan M R, Abdeltawab A A, et al. Sono – sorption efficiencies and equilibrium removal of triphenylmethane (crystal violet) dye from aqueous solution by activated charcoal [J]. Journal of Cleaner Production, 2019, 234: 124 – 131.

[8] Siao C – W, Chen H – L, Chen L – W, et al. Controlled hydrothermal synthesis of bismuth oxychloride/bismuth oxybromide/bismuth oxyiodide composites exhibiting visible – light photocatalytic degradation of 2 – hydroxybenzoic acid and crystal violet [J]. Journal of Colloid and Interface Science, 2018, 526: 322 – 336.

[9] Chen X, Shan X, Lan Q, et al. Electrochemiluminescence quenching sensor of a carboxylic carbon nanotubes modified glassy carbon electrode for detecting crystal violet based on nitrogen – doped graphene quantum dots@ peroxydisulfate system [J]. Analytical Sciences, 2019, 35 (8): 929 – 934.

[10] Marco – Brown J L, Guz L, Olivelli M S, et al. New insights on crystal violet dye adsorption on montmorillonite: Kinetics and surface complexes studies [J]. Chemical Engineering Journal, 2018, 333: 495 – 504.

[11] Jiang Y – R, Lin H – P, Chung W – H, et al. Controlled hydrothermal synthesis of BiOxCly/BiOmIn composites exhibiting visible – light photocatalytic degradation of crystal violet [J]. Journal of Hazardous Materials, 2015, 283: 787 – 805.

[12] Shen Y – D, Deng X – F, Xu Z – L, et al. Simultaneous determination of malachite green, brilliant green and crystal violet in grass carp tissues by a broad – specificity indirect competitive enzyme – linked immunosorbent assay [J]. Analytica Chimica Acta, 2011, 707 (1): 148 – 154.

[13] Hu Y, Gao Z. Yellow emissive Se, N – codoped carbon dots toward sensitive fluores-

cence assay of crystal violet [J]. Journal of Hazardous Materials, 2020, 388: 122073.

[14] Maxwell E J, Tong W G. Sensitive detection of malachite green and crystal violet by nonlinear laser wave mixing and capillary electrophoresis [J]. Journal of Chromatography B, 2016, 1020: 29 - 35.

[15] Zhou J, Qing M, Ling Y, et al. Double - stranded DNA nanobridge enhanced fluorescence of crystal violet/G - quadruplex complex for detection of lead ions and crystal violet [J]. Sensors and Actuators B: Chemical, 2021, 340: 129968.

[16] Zhu K, Hong Z, Kang S - Z, et al. Assembly of potassium niobate nanosheets/silver oxide composite films with good SERS performance towards crystal violet detection [J]. Journal of Physics and Chemistry of Solids, 2018, 115: 69 - 74.

[17] Roy S, Mohd - Naim N F, Safavieh M, et al. Colorimetric nucleic acid detection on paper microchip using loop mediated isothermal amplification and crystal violet dye [J]. ACS Sensors, 2017, 2 (11): 1713 - 1720.

[18] Lan Q, Li Q, Zhang X, et al. A novel electrochemiluminescence system of CuS film and K2S2O8 for determination of crystal violet [J]. Journal of Electroanalytical Chemistry, 2018, 810: 216 - 221.

[19] Yi K. A novel method for the quantitative determination of crystal violet utilizing YVO_4: Eu^{3+} nanoparticles [J]. Chemistry Letters, 2017, 46 (4): 520 - 523.

[20] Han Y, Chen Y, Liu J, et al. Room - temperature synthesis of yellow - emitting fluorescent silicon nanoparticles for sensitive and selective determination of crystal violet in fish tissues [J]. Sensors and Actuators B: Chemical, 2018, 263: 508 - 516.

[21] Han M, Zhu S, Lu S, et al. Recent progress on the photocatalysis of carbon dots: Classification, mechanism and applications [J]. Nano Today, 2018, 19: 201 - 218.

[22] Li T, Shi W, Mao Q, et al. Regulating the photoluminescence of carbon dots via a green fluorine - doping derived surface - state - controlling strategy [J]. Journal of Materials Chemistry C, 2021, 9: 17357 - 17364.

[23] Hu C, Li M, Qiu J, et al. Design and fabrication of carbon dots for energy conversion and storage [J]. Chemical Society Reviews, 2019, 48 (8): 2315 - 2337.

[24] Liu M L, Chen B B, Li C M, et al. Carbon dots: synthesis, formation mechanism, fluorescence origin and sensing applications [J]. Green Chemistry, 2019, 21 (3): 449 - 471.

[25]Li T, Shi W, E S, et al. Green preparation of carbon dots with different surface states simultaneously at room temperature and their sensing applications [J]. Journal of Colloid and Interface Science, 2021, 591: 334-342.

[26]Li H, Ye H-G, Cheng R, et al. Red dual-emissive carbon dots for ratiometric sensing of veterinary drugs [J]. Journal of Luminescence, 2021, 236: 118092.

[27]Liu X, Li T, Hou Y, et al. Microwave synthesis of carbon dots with multi-response using denatured proteins as carbon source [J]. RSC Advances, 2016, 6 (14): 11711-11718.

[28]Liu X, Li T, Wu Q, et al. Carbon nanodots as a fluorescence sensor for rapid and sensitive detection of Cr(VI) and their multifunctional applications [J]. Talanta, 2017, 165: 216-222.

[29]Li T, E S, Wang J, et al. Regulating the properties of carbon dots via a solvent-involved molecule fusion strategy for improved sensing selectivity [J]. Analytica Chimica Acta, 2019, 1088: 107-115.

[30]Jiang K, Sun S, Zhang L, et al. Red, green, and blue luminescence by carbon dots: Full-color emission tuning and multicolor cellular imaging [J]. Angewandte Chemie International Edition, 2015, 54 (18): 5360-5363.

[31]Jiang K, Sun S, Zhang L, et al. Bright-yellow-emissive N-doped carbon dots: Preparation, cellular imaging, and bifunctionalsensing, [J]. ACS Applied Materials & Interfaces, 2015, 7: 23231-23238.

[32]Wang G, Guo Q, Chen D, et al. Facile and highly effective synthesis of controllable lattice sulfur-doped graphene quantum dots via hydrothermal treatment of durian [J]. ACS Applied Materials & Interfaces, 2018, 10 (6): 5750-5759.

[33]He C, Xu P, Zhang X, et al. The synthetic strategies, photoluminescence mechanisms and promising applications of carbon dots: Current state and future perspective [J]. Carbon, 2022, 186: 91-127.

[34]Zhang R, Chen W. Nitrogen-doped carbon quantum dots: Facile synthesis and application as a "turn-off" fluorescent probe for detection of Hg^{2+} ions [J]. Biosensors and Bioelectronics, 2014, 55: 83-90.

[35]Vedamalai M, Periasamy A P, Wang C-W, et al. Carbon nanodots prepared from o-phenylenediamine for sensing of Cu^{2+} ions in cells [J]. Nanoscale, 2014, 6 (21):

13119 - 13125.

[36] Zhang Y - L, Wang L, Zhang H - C, et al. Graphitic carbon quantum dots as a fluorescent sensing platform for highly efficient detection of Fe^{3+} ions [J]. RSC Advances, 2013, 3 (11): 3733 - 3738.

[37] Wee S S, Ng Y H, Ng S M. Synthesis of fluorescent carbon dots via simple acid hydrolysis of bovine serum albumin and its potential as sensitive sensing probe for lead (II) ions [J]. Talanta, 2013, 116: 71 - 76.

[38] Zong J, Yang X, Trinchi A, et al. Carbon dots as fluorescent probes for "off - on" detection of Cu^{2+} and l - cysteine in aqueous solution [J]. Biosensors and Bioelectronics, 2014, 51: 330 - 335.

[39] Zhao D, Liu X, Zhang Z, et al. Synthesis of multicolor carbon dots based on solvent control and its application in the detection of crystal violet. Nanomaterials, 2019.

[40] Dong Y, Li J, Zhang L. 3D hierarchical hollow microrod via in - situ growth 2D SnS nanoplates on MOF derived Co, N co - doped carbon rod for electrochemical sensing [J]. Sensors and Actuators B: Chemical, 2020, 303: 127208.

[41] Sun S, Zhang L, Jiang K, et al. Toward high - efficient red emissive carbon dots: Facile preparation, unique properties, and applications as multifunctional theranostic agents [J]. Chemistry of Materials, 2016, 28 (23): 8659 - 8668.

[42] Ding H, Yu S - B, Wei J - S, et al. Full - color light - emitting carbon dots with a surface - state - controlled luminescence mechanism [J]. ACS Nano, 2016, 10 (1): 484 - 491.

[43] Bandi R, Kannikanti H G, Dadigala R, et al. One step synthesis of hydrophobic carbon dots powder with solid state emission and application in rapid visualization of latent fingerprints [J]. Optical Materials, 2020, 109: 110349.

[44] Gao D, Zhang Y, Liu A, et al. Photoluminescence - tunable carbon dots from synergy effect of sulfur doping and water engineering [J]. Chemical Engineering Journal, 2020, 388: 124199.

[45] Gao J, Liu Y, Jiang B, et al. Phenylenediamine - Based Carbon Nanodots Alleviate Acute Kidney Injury via Preferential Renal Accumulation and Antioxidant Capacity [J]. ACS Applied Materials & Interfaces, 2020, 12 (28): 31745 - 31756.

[46] Hu Y, Yang Z, Lu X, et al. Facile synthesis of red dual - emissive carbon dots for ra-

tiometric fluorescence sensing and cellular imaging [J]. Nanoscale, 2020, 12 (9): 5494 - 5500.

[47] Miao X, Yan X, Qu D, et al. Red emissive sulfur, nitrogen codoped carbon dots and their application in ion detection and theraonostics [J]. ACS Applied Materials & Interfaces, 2017, 9: 18549 - 18556.

[48] Song W, Duan W, Liu Y, et al. Ratiometric detection of intracellular lysine and pH with one - pot synthesized dual emissive carbon dots [J]. Analytical Chemistry 2017, 89 (24): 13626 - 13633.

[49] Pan D, Zhang J, Li Z, et al. Hydrothermal route for cutting graphene sheets into blue - luminescent graphene quantum dots [J]. Advanced Material, 2010, 22 (6): 734 - 738.

[50] Li Y, Hu Y, Zhao Y, et al. An electrochemical avenue to green - luminescent graphene quantum dots as potential electron - acceptors for photovoltaics [J]. Advanced Materials, 2011, 23 (6): 776 - 780.

[51] Liu R, Wu D, Feng X, et al. Bottom - up fabrication of photoluminescent graphene quantum dots with uniform morphology [J]. Journal Of the American Chemical Society, 2011, 133 (39): 15221 - 15223.

[52] Dong Y, Pang H, Yang H B, et al. Carbon - based dots co - doped with nitrogen and sulfur for high quantum yield and excitation - independent emission [J]. Angewandte Chemie International Edition, 2013, 52 (30): 7800 - 7804.

[53] Huang B, Zhou X, Xue Z, et al. Quenching of the electrochemiluminescence of Ru$(bpy)_3^{2+}$/TPA by malachite green and crystal violet [J]. Talanta, 2013, 106: 174 - 180.

[54] Liu B, Jiang W, Wang H, et al. A Surface Enhanced Raman Scattering (SERS) microdroplet detector for trace levels of crystal violet [J]. Microchimica Acta, 2013, 180 (11): 997 - 1004.

[55] Meng Y, Jiao Y, Zhang Y, et al. One - step synthesis of red emission multifunctional carbon dots for label - free detection of berberine and curcumin and cell imaging [J]. Spectrochimica Acta Part A: Molecular and Biomolecular Spectroscopy, 2021, 251: 119432.

[56] Hu Y, Gao Z. Sewage sludge in microwave oven: A sustainable synthetic approach to-

ward carbon dots for fluorescent sensing of para - nitrophenol [J]. Journal of Hazardous Materials, 2020, 382: 121048.

[57] Ling Y, Li J X, Qu F, et al. Rapid fluorescence assay for Sudan dyes using polyethyleneimine - coated copper nanoclusters [J]. Microchimica Acta, 2014, 181 (9): 1069 - 1075.

第7章
结论与展望

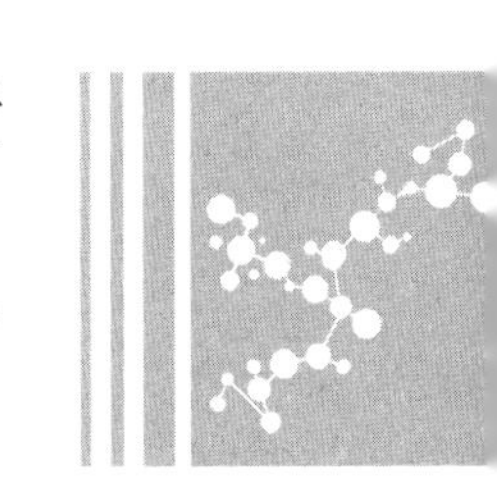

苯二胺衍生碳点的表面态调控策略及其
应用研究

7.1　结论

CDs结构的复杂性和多样性使其在荧光传感、生物成像及光学诊疗等领域展现了广泛应用前景。CDs是一种新型的碳纳米荧光粒子，由于其优异的理化性能使其有别于无机纳米晶体或有机荧光团等传统荧光材料，近年来备受关注。CDs的主要优点包括高的光稳定性、可调节的组成、易于制备/表面功能化和良好的生物相容性，这使它们成为纳米探针、催化、生物医学和光电器件领域的潜在材料。

本专著从分子融合角度出发，通过反应溶剂参与的分子融合法、室温氧化融合法、氟掺杂的表面态调控法以及溶剂热法制备了一系列不同表面态o-PD衍生CDs，并系统的研究了这些CDs的形成过程、发光机理，探索这些CDs在化学传感以及生物成像领域的应用。并得到以下结论：

(1)采用反应溶剂参与的分子融合法，设计制备了一系列不同表面态o-PD衍生CDs。其中以o-PD为碳源，甲酰胺为反应溶剂制备的FCDs为主要研究对象，提出了基于席夫碱反应的FCDs形成机理。研究发现，反应溶剂与前驱体分子之间的融合能够影响制备CDs表面态和结构，进而改善CDs的选择性。以FCDs为例，它可以通过荧光“开关”双模式作为探针用于Ag^+和Cys的检测。同时，基于“软碱”性质，FCDs对硬酸和临界酸类金属离子表现出了极强的耐受性。此外，FCDs具有长波长发射、良好的生物相容性和较低的细胞毒性，因此FCDs可以作为探针用于细胞成像，并能够防止活体组织自体荧光的干扰。最后，反应溶剂参与的分子融合法不仅为实用的功能性CDs设计提供了一条新颖的途径，而且为CDs的表面态/结构与性质之间关系的优化提供了新的见解。

(2)采用室温氧化融合法，以o-PD和HQ为原料，同时绿色制备了具有不同表面态的o-PD衍生CDs。在传统的CDs制备过程中，需要高温才能使有机小分子前驱体热解或碳化。而与之相比，该室温制备法具有操作简单、节省能源的优点，使CDs的制备成为真正的绿色工艺。基于制备的CDs不同的极性，通过硅胶柱层析将CDs进行分离，并对它们的荧光发射机理进行研究。同时，表面态的差异为CDs提供了的不同光谱特性使它们展现了不同的传感应用。其中，富含$-NO_2$和$-OH$基团的YCDs具有更大的波长发射，可以作为荧光探针用于有毒污染物p-NP检测。而富含$-NH_2$基团的GCDs则可以用于分析D_2O中H_2O的含量。此外，本工作通过室温氧化融合法绿色制备具有不同表面态CDs的方法极大地促进了制备多色/多用途CDs的研究进展。

(3)采用氟掺杂的表面态调控法,以 o - PD、4 - 氟 - 1,2 - 苯二胺和 p - BQ 为原料在室温下制备了不同氟含量的 CDs(FCDs1 和 FCDs2)。通过该方法制备的 CDs,当氟含量仅为 2.40% 时,CDs 的发射红移可以达到 70 nm。由于元素稀释效应和氢键的形成,FCDs1 中少量氟掺入可以缓解聚集诱导引发的荧光猝灭现象,并使其表现出固态荧光。随着氟含量的增加,由于浓度猝灭效应的发生,FCDs2 粉末的固态荧光消失。但是其溶液的荧光发射红移更多。最后,基于 FCDs1 固态荧光特性,将其用作 LFPs 显现的荧光可视化试剂。由于 FCDs1 较小的粒径,可以实现 LFPs 的高分辨率成像,荧光指纹显示出清晰的线条和明显的细节。而更高氟含量的 FCDs2,它的溶液荧光光谱与生物分子 CBL 的吸收光谱高度重叠,因此基于内滤效应将其作为荧光探针用于 CBL 的定量检测。本工作认为氟掺杂的表面态调控法的使用不仅为碳纳米材料的表面态/结构调控提供了一个有效的方法,而且为室温下制备所需荧光发射性能的 CDs 提供了新思路。

(4)采用简单的溶剂热法,以 o - PD 和乙醇为原料,制备出黄光氮掺杂 CDs(YNCDs)。基于内滤效应,姜黄素可以有效抑制 YNCDs 的荧光发射。因此,以 YNCDs 为纳米探针,建立了检测姜黄素的“开关”传感器平台,其检出限(3S/N)为 0.01 $\mu mol \cdot L^{-1}$。此外,该传感器平台实现了对咖喱粉、人尿、人血清等真实样品中姜黄素的高选择性、高灵敏度检测,准确性和回收率也令人满意。进一步地,YNCDs 优异的光学性能还可作为隐形油墨应用于信息存储和防伪等领域。本工作为 o - PD 衍生 CDs 的多领域应用提供了新方法。

(5)结晶紫是一种危险染料,对环境和人类健康构成严重威胁。这促使我们开发一种简便的方法来对其进行灵敏的检测。在此,我们建立了一种新型荧光纳米材料——强黄色碳点(SYCDs)对结晶紫的快速传感。以 o - PD 为前驱体,采用水热法制备了在 555 nm 处具有强发光 SYCDs。研究发现,SYCDs 的黄色荧光可以被结晶紫选择性猝灭,这使得 CDs 具有应用于结晶紫传感的可能。当结晶紫浓度在 0.02 ~ 15.00 $\mu mol \cdot L^{-1}$ 范围内线性良好,检出限低至 0.006 $\mu mol \cdot L^{-1}$。通过详细的研究,提出了内滤效应是传感机理。为实际应用,将新建立的方法应用于鱼组织样品中微量结晶紫的测定,取得了满意的结果。

7.2 展望

本专著以制备不同表面态 o - PD 衍生 CDs 为出发点,这些工作对于制备不同表面态 CDs 的思路起到了拓宽的作用。同时,对于认识 CDs 表面态/结构和性质/功能的关

系也具有积极意义。基于此,未来将对CDs的性质及应用进行更为深入的探究,主要涉及以下几个方向:

(1)相较于高温制备CDs的方法,室温制备CDs展现了极大的优势,但是有些荧光性能(例如:量子产率、红外发射等)仍然不如一些高温制备的CDs。因此,未来需要开发室温下制备高量子产率和近红外发射等更高荧光性能CDs的方法。

(2)目前,CDs发展面临的最根本和最主要的挑战之一,是缺乏一种理想的方法来制备具有理想结构的高质量CDs。换句话说,现有的方法不能准确地控制尺寸、形状、结晶度和表面态。同时,无论采用何种制备方法,得到的CDs都不可避免地含有多种杂质,如:小分子、聚合物、非晶态碳或碳颗粒等,这会对CDs的应用性能产生干扰,使CDs在某些特定应用中的作用机理不明确。此外,CDs的纯化需要复杂的后处理过程,包括透析、柱层析或电泳等,耗时长、成本高等问题严重阻碍了CDs的充分利用。因此,关于CDs制备过程中的监测和形成机理的解释有助于克服上述问题。

(3)室温制备CDs摆脱了传统高温制备CDs对于能量需求的限制。因此,在室温下如何大规模生产CDs也必将是未来的热门研究方向。

(4)本书所提出的几种制备不同表面态CDs的方法均是围绕o-PD作为碳源展开的,这其中必然存在一定的片面性。因此,未来需要开发更多的碳源,通过更为系统、更为深入的研究,更加全面地了解表面态对CDs性能的调控作用。

(5)目前,室温制备CDs需要较长的反应时间。因此,开发一些高效的催化剂从反应机理出发改进制备工艺、提高反应效率迫在眉睫。

(6)在生物成像和药物输送的应用中,探究水分散溶剂对CDs的光学性能的影响是一个不可避免的课题。同时,CDs在水中具有长波长(近红外)发射和较好的光学性质对于生物应用是十分必要的。与m-PD相比,o-PD和p-PD更多地被选作制备长波长发射CDs的碳源。基于o-PD制备的CDs的荧光发射一般在560 nm以内,这将限制此类CDs在生物领域的应用,而p-PD制备的CDs往往水溶性差,不利于生物应用。因此,未来需要开发新的策略来获得适用于生物领域的苯二胺衍生CDs。